DEVELOPMENTS IN STRESS ANALYSIS—1

THE DEVELOPMENTS SERIES

Developments in many fields of science and technology occur at such a pace that frequently there is a long delay before information about them becomes available and usually it is inconveniently scattered among several journals.

Developments Series books overcome these disadvantages by bringing together within one cover papers dealing with the latest trends and developments in a specific field of study and publishing them within *six months* of their being written.

Many subjects are covered by the series, including food science and technology, polymer science, civil and public health engineering, pressure vessels, composite materials, concrete, building science, petroleum technology, geology, etc.

Information on other titles in the series will gladly be sent on application to the publishers.

DEVELOPMENTS IN STRESS ANALYSIS—1

Edited by

G. S. HOLISTER

Faculty of Technology, The Open University,
Walton Hall, Milton Keynes, UK

APPLIED SCIENCE PUBLISHERS LTD
LONDON

APPLIED SCIENCE PUBLISHERS LTD
RIPPLE ROAD, BARKING, ESSEX, ENGLAND

British Library Cataloguing in Publication Data

Developments in stress analysis. 1 (Developments
series).
1
1. Strains and stresses
I. Holister, G S
620.1'123   TA407

ISBN 0-85334-812-X

WITH 6 TABLES AND 80 ILLUSTRATIONS

© APPLIED SCIENCE PUBLISHERS LTD 1979

Printed in Great Britain by Galliard (Printers) Ltd, Great Yarmouth

PREFACE

The techniques of experimental and numerical stress analysis lie at the heart of engineering practice. All man-made artifacts need to be designed and constructed in a manner which ensures that, in use, they do not fall apart. Structures and mechanisms that are used by the general public have to be constructed with safety and reliability as the over-riding factor. Yet considerations of weight, convenience and cost also require the engineer to optimise his designs in a manner which enables the manufacturer to 'get more out of less'—Buckminster Fuller's well-known definition of engineering.

This cannot be done on the drawing board alone. A load-bearing structure will be designed to sustain a specified load safely. Knowing the maximum loads, the deformation and hence the strains suffered by the structure can then be calculated, and the structure designed so that all its component parts can safely support such strains. Prototypes or scale models must then be constructed and tested to check the accuracy of the design. At this stage the stress analyst is responsible for ensuring the safety and structural integrity of the prototype. If he errs, invested capital, and perhaps even lives, may be lost. When the final structure is eventually put into service the stress analyst still has an important function in monitoring stresses and strains in the working components of the structure, and, in the event of failure due to poor design, material defects, or unforeseen circumstances, he will be at the heart of the work required to modify the structure to permit its return to service.

Stress analysis is, therefore, a vital function which is central to technological activity. Like all technology, its techniques and tools are constantly being modified and improved, and consequently there is an urgent need for sources to which the practising engineer can turn to ensure that, in facing a given problem, he is using the best available technique.

We hope that this book, which describes the latest 'state of the art' in the most commonly used stress analysis methods available to us, will fulfil that function.

G. S. HOLISTER

CONTENTS

LIST OF CONTRIBUTORS

A. R. LUXMOORE
Department of Civil Engineering, University College of Swansea, Swansea SA2 8PP, UK.

P. S. THEOCARIS
The National Technical University, 5 Zographou Street, Athens 625, Greece.

A. L. WINDOW
Welwyn Strain Measurement Ltd, Armstrong Road, Basingstoke, Hampshire RG24 0QA, UK.

O. C. ZIENKIEWICZ
Department of Civil Engineering, University College of Swansea, Swansea SA2 8PP, UK.

Chapter 1

NUMERICAL METHODS IN STRESS ANALYSIS—
THE BASIS AND SOME RECENT PATHS OF
DEVELOPMENT

O. C. ZIENKIEWICZ

University College, Swansea, UK

SUMMARY

This chapter deals in a general way with the application of numerical methods in stress analysis. Although the basic outline of finite differences and finite element methods is given elsewhere a brief introduction to the subject is made here and an attempt is presented to set the stage for recent developments which have already taken place and others where the research is just starting. In this context possibilities offered by boundary solution technique, finite element methods using reduced integration, etc., are discussed.

INTRODUCTION

The mathematical theory of elasticity and its governing equations date back to the beginnings of the nineteenth century—and little has been added to the basic mathematical model since. The exact solution of these equations presents however insuperable difficulties in all realistic situations and since the start of this century much effort has been devoted to devising practical means for their approximate, numerical, solution. Today with the use of powerful digital computers such numerical solutions are readily available for most complex problems of elasticity—and indeed are superseding the alternative of experimental modelling processes due to the ease of changing design parameters, material properties, etc.

The more recent interest in the inclusion of more complex nonlinear material relationships such as creep, plasticity, etc. has made the possibility of exact solutions even more remote and the adaptability of numerical procedures for treatment of such models makes them even more popular in this area. Here, however, the difficulty of describing fully and realistically the material properties retains the need for appropriate physical experiments and this phase of numerical stress analysis is much dependent on experimental corroboration. Nevertheless, once again numerical solutions present possibilities difficult or impracticable to realise in the laboratory. For instance reproduction of 30 or more years of the working life of a nuclear reactor in a relatively short computer run is but one of such examples.

In this chapter we shall consider briefly the essence of numerical approximation processes used—and attempt to present a picture of the present state of the art.

Historically three, apparently different, processes of approximation to problems of stress analysis (or more generally to that of solving boundary value problems defined by certain governing partial differential equations) have developed almost in parallel. These are: (a) finite difference methods, (b) finite element methods—with their ancestry of Rayleigh–Ritz approximations and (c) boundary solution methods.

Today each methodology is still practised and each has its more or less violent protagonists or opponents. Further, more rationally, each formulation has its specific merits and limitations, and apparently its own field of optimal applicability.

It is the author's view that all such methodologies fall within a single class of processes which may be called *the generalised finite element–trial function method* and the understanding of this fact permits the universally optimal procedures to be applied without personal prejudices. We shall attempt to present in the next section a brief outline of the mathematical basis of approximation involved in each process keeping within the space limitations of the chapter. Clearly much additional reading will be necessary for a fuller understanding of each method and here the reader is referred to modern texts in the subject. Though such texts are selected as representative of each type of methodology respectively[1-3] a brief historical review is appropriate at this stage.

The *finite difference method* is probably the first process by which elasticity problems with arbitrary boundaries have been solved. The classical work of L. F. Richardson of 1910[4] pioneered the way—albeit with considerable computational difficulties.† Southwell's relaxation methods[5] reduce considerably the solution effort yielding finally to computer based calculations.

The computer area, which was entered into in the early 1950s, permitted structural problems with many discrete elements to be solved simultaneously.[6] It was natural therefore to attempt to model the elastic continuum as an assemblage of *'finite elements'* physically approximating to it.[7]

From such a modest beginning it was soon realised that the essence of the finite element method lay in the trial function approximations of Rayleigh,[8] Ritz,[9] Galerkin,[10] and others using now, however, somewhat special trial function forms. This general basis gave to the process an aura of respectability and led to developments which today make the method so widely applicable that a bibliography of references[11] shows an annual rate of publication of 1400 in 1974 and the rate is still increasing.

The *boundary solution method* appears to have been first conceived by Trefftz[12] and although used extensively was first presented as a practical computational tool of stress analysis by Massonnet.[13] The work of Rizzo,[14] Cruse,[15] Butterfield,[16] and others has led to economical solution methods of homogeneous and linear elastic situations and this procedure is becoming more widely used today. The links of this process with the finite element method are close even if at first not obvious.[17,18]

THE MATHEMATICAL BASIS OF NUMERICAL SOLUTIONS

The Specific Example
To illustrate the alternative means of numerical discretisation (and solution) processes we shall focus attention on one specific differential equation corresponding to that of elastic torsion of prismatic bars. Here we shall consider the solution of the following differential equation[4]

$$\frac{\partial^2}{\partial x^2}\psi + \frac{\partial^2 \psi}{\partial y^2} \equiv \nabla^2 \psi = 0 \qquad (1a)$$

† To quote Richardson, 'So far I have paid piece rates for the operation of about n_{18} pence per coordinate point, n being the number of digits. The chief trouble to the computers has been the intermixture of plus and minus signs. As to the rate of working, one of the quickest boys averaged 2000 operations per week for numbers of three digits, those done wrong being discounted.' Even accounting for rates of inflation this is a rather expensive operation compared with the cost of present-day machines.

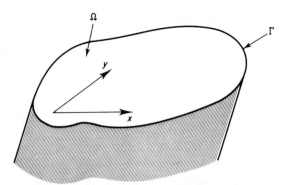

FIG. 1. Torsion problem—its domain Ω and boundary Γ.

in the two-dimensional region Ω subject to prescribed values

$$\frac{\partial \psi}{\partial n} = \bar{q} \tag{1b}$$

on the boundary (Fig. 1).

ψ is known as the warping function.†

While this simple problem is probably the easiest one a stress analyst is called upon to tackle, the basic steps of analysis can be here very readily followed.

The Finite Difference Process

The object of discrete numerical analysis is to convert the continuum problem given by eqns. (1) into a set of algebraic relations which can be solved by a computer or by hand calculator. The apparently most obvious procedure is to replace the derivatives of the function ψ in eqn. (1) by suitable approximations using a set of nodal values of the unknown function. Thus for instance consider a typical point 0 of a regular mesh of points shown in Fig. 2. We could write for instance

$$\left(\frac{\partial \psi}{\partial x}\right)_0 \approx (\psi_1 - \psi_3)/2h$$

$$\left(\frac{\partial^2 \psi}{\partial x^2}\right)_0 \approx (\psi_1 - 2\psi_0 + \psi_3)/h^2 \tag{2}$$

† N.B. This formulation is not in fact often used with the stress function alternative being more popular. It is however more illustrative in the present problem.

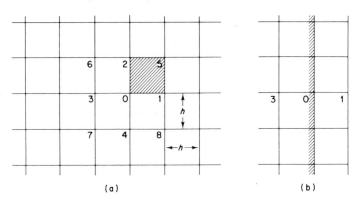

FIG. 2.   A finite difference mesh; (a) interior, (b) boundary with gradient condition.

as the lowest possible approximation to the first and second derivatives.
Using these we can replace eqn. (1) by its finite difference equivalent;

$$\psi_1 + \psi_2 + \psi_3 + \psi_4 - 4\psi_0 = 0 \qquad (3)$$

Further, if values of ψ are specified at the boundaries immediately we
could write the equation system in matrix form as

$$\mathbf{K}\boldsymbol{\psi} = \mathbf{c} \qquad (4)$$

where the vector

$$\boldsymbol{\psi} = [\psi_1, \psi_2 \ldots \psi_n]^\Gamma \qquad (5)$$

lists all the unknowns and $\mathbf{c}$ is obtained by substitution of known values of ψ
on the boundaries and $\mathbf{K}$ is a symmetric non-singular square matrix.

If, as in the case discussed, the gradient of ψ, i.e. $\partial\psi/\partial n$ is specified on the
boundaries then the equations of nodal points on boundaries have to be
supplemented by additional ones casting eqn. (1b) in a finite difference
form. For a typical boundary point of a straight boundary shown in Fig.
2(b) we thus could write

$$(\psi_1 - \psi_3)/2h = \bar{q} \qquad (6)$$

as an additional equation.

Once again the resulting equation system for the whole unknown vector
ψ can be cast into form (4) but now in general $\mathbf{K}$ is no longer a symmetric
matrix as the reader can easily verify. This lack of symmetry does not
present serious solution difficulties in the finite difference scheme which
generally for a regular mesh is carried out in an *iterative manner*.[1] However,

it is a point to be noted as we shall observe later that this need not arise in an almost identical finite element formulation.

While the derivation of the finite difference equations in the above example was simple, it presents considerable difficulties if:

(a) an irregular mesh is used
(b) higher order finite difference approximations are to be introduced in place of those in eqn. (2)
(c) material inhomogeneities and in particular discontinuous 'jumps' of material properties occur.

The reader should note that the matrices **K** will always be narrowly interconnected (banded) which facilitates the solution.

The Finite Element Method

The basis of the method is to consider now not the differential equation but its *weighted form*. Thus in place of eqns. (1a) and (1b) we can write with complete validity

$$\int_{\Omega} W\left(\frac{\partial^2 \psi}{\partial x^2} + \frac{\partial^2 \psi}{\partial y^2}\right) dx\, dy - \int W\left(\frac{\partial \psi}{\partial n} - \bar{q}\right) d\Gamma = 0 \qquad (7)$$

where W is any (mathematically admissible) function. Clearly eqns. (1a) and (1b) or (7) are equivalent. An integration by parts (or the use of Green's identity) for the first integral allows us to write the above in the alternative (weak) form simply as

$$\int_{\Omega} \left(\frac{\partial W}{\partial x} \cdot \frac{\partial \psi}{\partial x} + \frac{\partial W}{\partial y} \cdot \frac{\partial \psi}{\partial y}\right) dx\, dy + \int_{\Gamma} W\bar{q}\, d\Gamma = 0 \qquad (8)$$

and this will be found to be particularly useful.

If now the unknown function ψ is expanded as an approximation in a series of prescribed trial functions $N_i(x,y)$, as

$$\psi = \sum_{i=1}^{n} N_i Q_i = \mathbf{NQ} \qquad N_i = N_i(x,y) \qquad (9)$$

and further a substitution made into eqn. (7), then the bracketed terms represent simply the residual (or error) in satisfaction of eqn. (1). Clearly this error has to be minimised in some manner and one way of ensuring this is to admit that eqn. (7) or eqn. (8) is satisfied exactly only for, n, different but specific choices of the weighting function $W = W_i$, $i = 1 - n$. The general methodology is known as the method of weighted residuals[19,20] and once again a simple algebraic system of equation results in a form

$$\mathbf{KQ} = \mathbf{C} \tag{10}$$

If the start is made from eqn. (8) then it is easy to show that in a general form the matrix coefficients are

$$K_{ij} = \int_{\Omega} \left(\frac{\partial W_i}{\partial x} \cdot \frac{\partial N_j}{\partial x} + \frac{\partial W_i}{\partial y} \cdot \frac{\partial N_j}{\partial y} \right) dx \, dy \tag{11a}$$

$$C_i = \int_{\Gamma} W_i \bar{q} \, d\Gamma \tag{11b}$$

Many choices are obviously possible for both sets of functions N_i and W_i and their relation. For instance if we take

$$N_i = W_i \tag{12}$$

the method becomes one associated with the name of Galerkin[10] and indeed results in specific cases in precisely the well known energy formulations of Rayleigh[8] and Ritz.[9]

In the classical approaches of the precomputer era the functions were usually defined continuously by a form of trigonometric Fourier expansion over the whole domain. Clearly, however, this is not essential as any integral can be written in an additive way. Thus we can define all functions N_i (and W_i) separately for various subregions or elements and we can write

$$K_{ij} = \sum_{e=1}^{m} K_{ij}^e \qquad C_i = \sum_{e=1}^{m} C_i^e \tag{13}$$

where expressions (11) serve to evaluate the now appropriate element contributions. We note immediately that for the Galerkin process applied to the above example the equations will always be fully symmetric—a considerable advantage over the finite difference form.

If we now identify the general parameters $\mathbf{Q}$ with nodal values of the function ψ, the approximation takes the form of eqn. (4). To be specific, consider for instance that each rectangle in the mesh of Fig. 2(a) represents an element and that the shape functions are defined by bilinear expansion shown in Fig. 3 giving for an element '0152'

$$N_0 = (x - h)(y - h)/h^2$$
$$N_2 = (x - h)y/h^2, \text{ etc.} \tag{14}$$

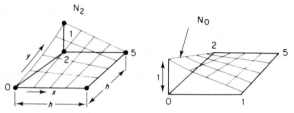

FIG. 3.   Square finite element and bilinear shape functions N_2 and N_0.

A little computation shows that a typical assembled equation for a node such as 0 in Fig. 2(a) becomes now

$$\psi_1 + \psi_2 + \cdots \psi_8 - 8\psi_0 = 0 \qquad (15)$$

This expression is similar but not identical to that pertaining to the simple finite difference approximation of eqn. (4).

Consider however a set of triangular subdivisions as shown in Fig. 4 in which again a linear interpolation function shown there is used. A calculation of a similar kind to the above shows that in this case precisely the same equation as that given by the simple finite difference algorithm is achieved.[2]

The reproduction of the finite difference algorithm by a particular finite element pattern and type is quite general and occurs in a variety of problems. We can therefore conclude that the finite difference algorithm is but a subclass of the more general trial function approach which in addition has the following merits:

(a)   By suitable (Galerkin) approximation the boundary conditions can be incorporated into the main formulation yielding a similar accuracy throughout the problem. (We note here that the local finite difference boundary approximation typified by eqn. (2) not

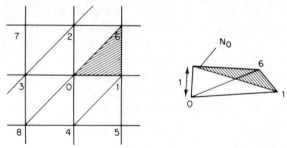

FIG. 4.   A triangular element subdivision of a regular mesh.

only leads to non-symmetric equation but is of lower order approximation than that inherent in eqn. (3).)

(b) Arbitrary shape of elements can be used without changing the basic algorithm.

(c) Material discontinuities can be readily incorporated without any additional approximation or elaboration of element characteristics.

(d) If necessary higher order elements can be used with more complex interpolation patterns without difficulty and thus a higher order of approximation achieved.

While the computation of nodal equations is obviously more difficult if done explicitly, the computer assembly is very rapid. If attention were restricted to regular meshes then indeed even this step could be avoided once a typical equation is computed and even the small merit of finite difference algorithms now disappears.

While many varieties of finite elements have been developed, some of these applicable to the problems in hand being shown in Fig. 5, on many

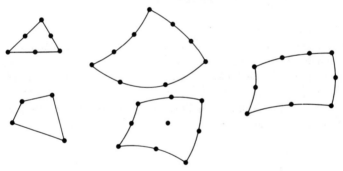

Fig. 5. Some more complex elements.

occasions (particularly in dynamic problems) the use of simple elements is preferred. Thus we can return here to algorithms almost identical to those of finite difference and these may here be only superseded in name and perhaps philosophy of approach rather than in hard computational fact.

Boundary Solution Methods

In this process once again trial function solutions are used albeit not always explicitly. Expansions in the form of eqn. (9) are still basically valid.

$$\psi = \sum N_i Q_i \equiv \mathbf{NQ} \tag{16}$$

but now the shape functions N_i are so chosen that the governing equations (1a) are automatically satisfied, i.e.

$$\nabla^2 N_i = 0 \qquad (17)$$

Numerically all that remains to be done is to achieve the satisfaction of the boundary conditions, e.g. eqn. (1b) in an approximate sense. Once again either a pointwise satisfaction is attempted or more generally, just as in the finite element analysis, a weighted form of approximation is used. The latter follows immediately from eqn. (7) if we note that the first term is identically zero. Thus we can write for the present case

$$\int_\Gamma W \left(\frac{\partial \psi}{\partial n} - q \right) d\Gamma = 0 \qquad (18)$$

Clearly the method of approximation now used has some different characteristics and in particular:

(a)   Fewer discrete parameters need to be used as approximation is confined only to the boundary where a discretised equation of the following form is achieved

$$\mathbf{KQ} = \mathbf{C} \qquad (19)$$

(b)   The structure of the matrix $\mathbf{K}$ is now not banded as each function N_i will affect all the boundary points.

(c)   It is easy to deal with infinite domains or singularities as analytical functions for such are readily found.

(d)   The method is limited to linear and homogeneous regions for which analytical solutions exist.

The determination of trial functions N_i can here follow several paths. In the most obvious case an intelligent selection of suitable functions designed to suit the problem in hand is made. Such procedures are typical of the edge function method developed by Quinlan.[21]

In an alternative to this a *distribution* of sources (or point singular functions) is taken along the boundary and approximated by a finite set of parameters. Such a procedure guarantees completeness and therefore convergence can be more readily discussed and the method has been used in potential problems by Von Karman[22] and more recently the solution of elasticity problems.[13–16,23]

Procedures described above are known as *indirect* and can readily be combined to achieve the merits of both.[24,25] Indeed combination can be made with the finite element method in a simple way.

The third type of boundary solution uses as its basis an integral equation in which no auxiliary variables arise. Thus the discretisation is achieved with ψ and $\partial\psi/\partial n$ specified at some boundary nodes. The basis of such approximations is nevertheless identical.[25] Much work of recent years has been derived by this approach.[14,15,26]

GENERAL DESCRIPTIONS OF NUMERICAL PROCEDURES— THE PRESENT AND THE FUTURE

In the preceding section we have outlined three general procedures of numerical solution applicable to stress analysis problems. Although the example concerned a rather simple homogeneous and linear torsion case, similar general procedures are applicable to all continuum problems. Indeed they can be generalised to non-linear situations in two and three dimensional space. If dynamic problems arise the discretisation allows in a similar manner to reduce the partial differential equations to ordinary ones with time remaining as the only independent variable. The final equation can then once again be solved using any of the three numerical procedures described (here finite difference methods are still most popular) and indeed exact solutions can often be obtained. We can at this stage therefore summarise the merits and disadvantages of the various techniques for all cases.

1. The finite element method is undoubtedly the most versatile and widely applicable one. Many variants in the choice of basic parameters shape and weighting functions exist and with some of these the algorithms of finite difference techniques are reproduced identically. We have no reason therefore to discuss the latter separately.
2. The boundary solution methods are more limited in applicability but in linear homogeneous domains (especially if these contain singularities or arc of infinite extent) can provide a most economical tool.

The first of the propositions is now widely recognised though perhaps not by the most rigid adherents of the finite difference processes.† The great

---

† In some more recent so called 'finite difference' forms the procedure is often started from integral statements, i.e. equations of the type given by (7) and (8). With such formulations even conceptual differences between the finite element and the finite difference method disappear. The reader can consult here the work of Varga[27] Griffin and Kellog[28] and Wilkins.[29]

versatility of the finite element method possesses in itself however some seeds of difficulty. Which element or which formulation of the many available is best to use in a given situation? Will new formulations supersede those in current use?

Here the answer is not simply noting that we can approach a given problem of stress analysis with many variants. We can therefore:

(1)   Choose displacement, stresses, stress functions, etc., as dependent variables.
(2)   Use different forms of shape function approximation (N_i).
(3)   Use different weighting functions (W_i).
(4)   Noting that an approximation is introduced by the choice of the expansion we can
    (a) add to this by using an additional approximation in evaluation of integrals such as that given by the terms of eqn. (11).
    (b) Violate certain continuity requirements[2] of expansion or impose these by additional variables.

Clearly the user needs an assurance that the process is convergent (i.e. that as element subdivision is made finer exact answers are approached). Further he needs a knowledge of cost effectiveness, i.e. of the accuracy obtainable with a given subdivision and element. He will have to decide which program and which element will be cheapest to achieve the accuracy desired.

Today the questions of convergence can be answered simply by the so called patch test[2,30,31] but the questions of error estimates are still much more in dispute in the research stage. Here the recent advent of mathematicians of the finite element stage is much to be welcomed.[32] Already some significant steps have been taken[33,34] and we may well be approaching the stage where the user of a finite element program can be assured that for his particular subdivision an error bound (in stress or displacement) can be estimated.

The final question of element choice still remains. Here what is the best may never be determined and such factors as efficiency of the program, the method used in it for solving the discrete equations (elimination or iteration), etc., will always arise. It is now well established that simple conforming elements with displacement variables are not the most accurate for a given subdivision. So called hybrid elements,[2,35] and, more simply, elements using non-conforming expansions or inexact numerical integration are better performers. A wider use of such approximations is to be anticipated in the future.

We have noted from the example of Fig. 5 that elements with a large number of nodes (or parameters) exist today. Are these preferable to simple ones? Here we shall first disregard one of the arguments used in the past for such complex elements, i.e. that of data preparation. This will be discussed later but we note here in passing that with modern computers, input can be generated automatically and it is equally easy to generate a large number of simple elements as a smaller number of more complex ones. Thus for a given number of nodes (or degrees of freedom) we find that undoubtedly the complex element is more accurate. Figure 6, for instance, illustrates two somewhat similar nuclear reactors for which stress analysis with simple tetrahedra and the more complex isoparametric bricks were used. In the first, some fifteen thousand degrees of freedom were involved, while in the latter only twelve hundred for half of the assembly.[2,36] Clearly the latter is more 'economical' if the elimination type of solution is used as is the practice in the majority of present day computer programs. However, as three dimensional analyses are costly due to the very wide banding interconnection there is much to be said for the possible use of iterative solution procedures which have been extensively developed in the finite difference field but not used widely with finite elements. With such solution methods the banding is immaterial but a simple interconnection is a positive advantage. It is more than likely therefore that with programs in which such iterative solutions are adopted the cost advantage, possibly even for problems as complex as those illustrated in Fig. 6, will be reversed and simple elements will become more desirable. This indeed may be crucial as progressively larger and non-linear problems are being tackled in which a large number of linear solutions are necessary. An indication of this trend is already discernible in dynamics where the step by step solution processes in time are very similar to those adopted in the iterative solution of non-linear problems.[27] Recent work suggests that considerable economy is achieved by use of explicit solutions here as these do not involve the solution of simultaneous equations at every step.[37,38] In Figs. 7(a)–(e) we show such a dynamic solution using the problem of stress and deformation determination of an earthquake response for a typical earth dam (see also Table 1). The two meshes shown achieve similar accuracy, although the number of nodes with the complex (quadratic) element is much smaller than that used with simpler (linear) element, the solution time is much shorter with the latter.

At this stage it is worth remarking that the solution of static problems may well be effectively approached in an identical manner to that used in dynamics but with artificial use of mass and damping characteristics. This

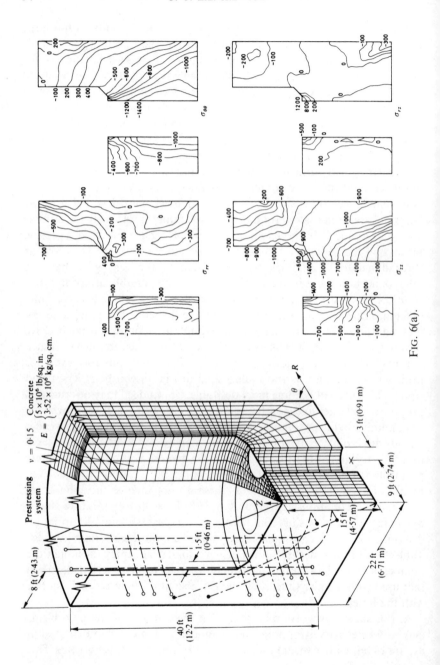

Fig. 6(a).

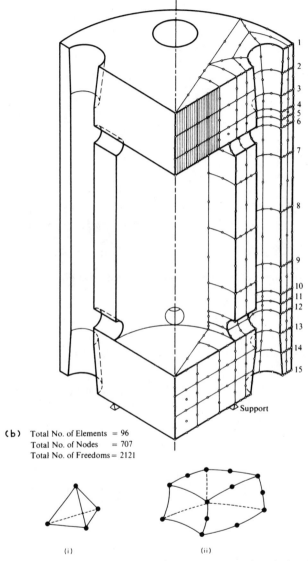

(b) Total No. of Elements = 96
     Total No. of Nodes   = 707
     Total No. of Freedoms = 2121

(i)   (ii)

FIG. 6.   Pressure vessel analyses. (a) A nuclear pressure vessel analysis using simple
tetrahedral elements. Geometry, subdivision and some stress results. (i) Basic
element: tetrahedron with 12 degrees of freedom. (b) Three-dimensional analysis of
a pressure vessel using isoparametric hexahedra. (ii) Basic element: curvilinear
hexahedron with 60 degrees of freedom. Copyright © 1977, McGraw-Hill, New
York.

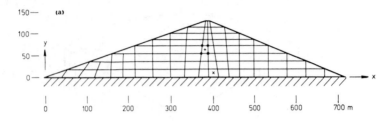

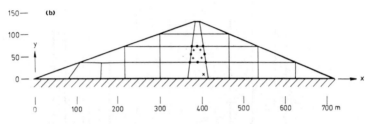

FIG. 7.   (a) Finite element mesh of 4 node elements. (b) Finite element mesh of 8 node elements.

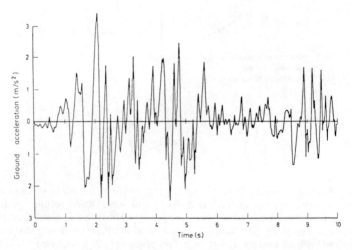

FIG. 7.   (c) Acceleration input.

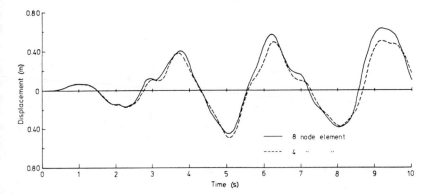

FIG. 7. (d) Horizontal displacement response.

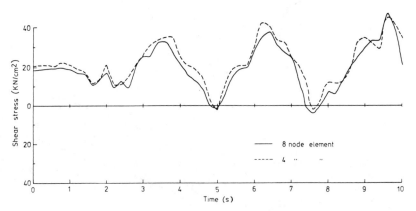

FIG. 7. (e) Shear stress response.

TABLE 1
DETAILS OF ANALYSIS

Type	Nodes	Elements	Degrees of freedom	Gauss points	Time step length (s)	Computation time (s)	Comments
4 node element	122	92	244	92	0·012	56	cheaper
8 node element	102	25	204	100	0·006	133	more accurate

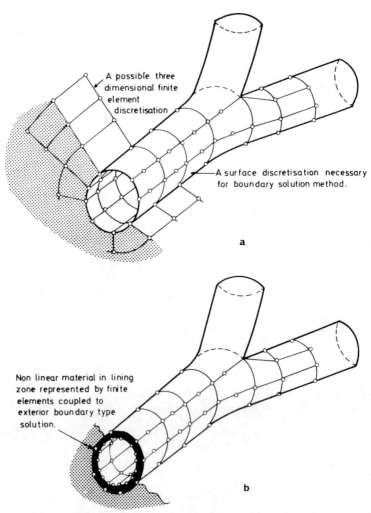

FIG. 8.   Problem illustrating merits of boundary solution processes and possible use of coupling with finite elements. Analysis of a tunnel in an infinite elastic and homogeneous medium with inelastic lining.

form of calculation is the basis of so called 'dynamic relaxation' techniques first developed in the finite difference context.[39-41]

Such methodologies could well be used in the next program generation in a similar manner to that adopted in various existing finite element codes but now adding the generality of a finite element approach. Indeed the

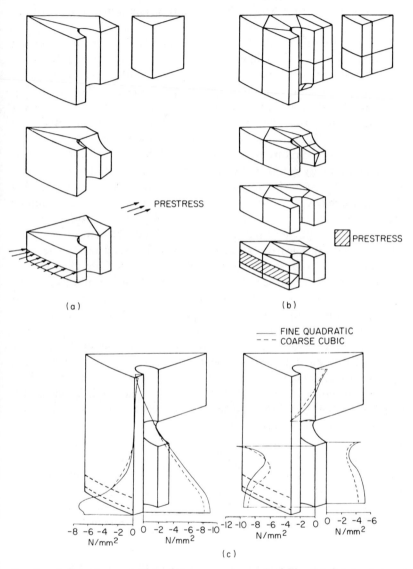

FIG. 9. A boundary solution of the pressure vessel of Fig. 6 (after Lachat and Watson[26]); (a) coarse subdivisions, (b) fine subdivisions, (c) circumferential stress due to prestress. Reproduced with permission from *Int. J. Num. Meth. Eng.*, **10,** pp. 1001–3. Copyright © 1976, John Wiley & Sons Ltd, New York.

similarity of these processes with formal iteration has been fully discussed in mathematical literature by Varga.[27]

Again in the same vein the use of an extremely effective technique developed for the finite difference processes and not suffering from the inherent numerical instability of explicit methods could well be adopted for finite elements in the future. This technique known as the 'alternating direction implicit' or ADI method[1] is well known. All such developments of computation may well affect the problem of 'best element' and make the present choices largely subjective and program-dependent.

Finally we should return to the boundary solution process and its merits. As we have already mentioned this is extremely efficient in homogeneous and linear situations. Figure 8(a) shows for instance an example of a problem where due to the small volume-to-boundary ratio and the infinite extent of the domain the boundary solution method would be very economic. The finite element discretisation of this problem would certainly not be competitive. Clearly for such problems we foresee a development of a series of special programs competing effectively with finite elements but even here the choice presents many difficulties. Consider once again the problem of Fig. 6 in which rather complex and slightly non-homogeneous though linear domain is present. The boundary solution method presents here a difficulty owing to the very wide banding if a straightforward discretisation were adopted, and to overcome this a series of subdomains, very similar to standard finite elements (Fig. 9) is used in practice.[26] The circle is thus beginning to close and we could consider a 'marriage' of the finite element concept with the boundary solution methods in a manner that would allow the use of the best feature of each process to be obtained. Outlines of the problems presented by such a marriage are given in references 17, 18, 24 and 25 and once again this will probably occur most when adopting the finite element assembly methodology.

In Fig. 8(b) is shown a situation where, for instance, a linear exterior solution of a tunnel problem can be combined with a non-linear behaviour of a lining for which standard finite elements are most efficient. Clearly here the merits of both methods could be successfully coupled.

LOGISTICS OF NUMERICAL STRESS ANALYSIS

In the preceding section we have discussed the essential basis and inherent merits of the various possible approaches. From a practical point of view such features as ease of data generation, output visualisation, etc., are often the controlling factors in overall economy. Thus, while the understanding of the basis of all methods is essential for any intelligent use of a given program or method, it is extremely important that good facilities should be available for input and output. Indeed it is often noticeable today that some programs using obsolete methodology are more popular than others more advanced due to good logistics of this kind.

The subject is a wide one involving computer scientists working in parallel with engineers and serious development effort in this area has existed and continues to exist.

It is important to realise here such problems as:

(a)   The ease of geometrical and loading specification of the problem. Here a certain minimum of user input can obviously not be avoided.
(b)   The ease of internal subdivision generation necessary for numerical analysis. This part ideally in some distant future is going to be made completely automatic and based on error estimates and other needs of the solution process. However, at this stage, control over discretisation is needed but should be exercised in a most simple manner.
(c)   The realisation that often identical techniques (i.e. FEM) are used for the solution of loading or response parts of the problem which may be entirely of non-structural nature. Typical here are methods of coupling external fluid pressures[42] and temperature computation. With such parallel computations it is essential that ease of transfer data for each phase of solution should be available.
(d)   The output of stress computation, unlike that in experimental stress analysis, often produces much more information than immediately required so the storage of potentially useful results and of its graphic output is essential.

In the context of the present article it is impracticable to discuss in any detail each of these phases and reference to some recent development in each phase is given. Thus, for geometrical input and mesh generation the reader may well consult the references.[42,43] For output presentation many papers on computer graphics are now available.

In reference 44 a simple perspective manner of showing for instance three-dimensional stress analysis results is indicated and this procedure has much merit (Fig. 10). Obviously other alternatives for two- and three-dimensional stress presentation exist and are being developed with the advent of better black and white, and colour display possibilities.

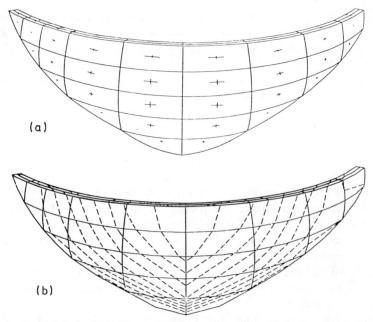

(a)

(b)

FIG. 10.   Representation of three-dimensional stresses. Upstream face of arch dam; (a) vector stress plot (b) contours of smaller principal stress, eye point $x = -600$, $y = 0$, $z = -200$, centre vision $x = y = z = 0$.

CONCLUDING REMARKS

In the short space of this article only a sketchy reference to many important new developments has been made. For instance we have not talked about 'global element'[45] or 'infinite element'[46] concepts which are finding an ever increasing applicability. Other particular and important steps forward have been omitted but the objective of this short survey will be achieved if the major fronts of recent activity are noted.

REFERENCES

1.(a) MITCHELL, A. R., *Computational Methods in Partial Differential Equations*, Wiley, New York, 1969.
   (b) MITCHELL, A. R. and WAIT, R., *The Finite Element Method in Partial Differential Equations*, Wiley, New York, 1977.

2. ZIENKIEWICZ, O. C., *The Finite Element Method*, 3rd edition, McGraw-Hill, London and New York, 1977.
3. JAWSON, M. A. and SYMM, G. T., *Integral Equation Methods in Potential Theory and Elastostatics*, Academic Press, London and New York, 1977.
4. RICHARDSON, L. F., The approximate arithmetical solution by finite differences of physical problems. *Trans. Roy. Soc. (London)*, 1910, **A210**, 307–57.
5. SOUTHWELL, R. V., *Relaxation Methods in Theoretical Physics*, Clarendon Press, Oxford, 1946.
6. ARGYRIS, J. H., *Energy Theorems and Structural Analysis*, Butterworth, London, 1960. (Reprinted from Aircraft Eng., 1954–55.)
7. TURNER, M. J., CLOUGH, R. W., MARTIN, H. C. and TOPP, L. J., Stiffness and deflection analysis of complex structures, *J. Aero. Sci.* 1956, **23**, 803–23.
8. LORD RAYLEIGH (STRUTT, J. W.), On the theory of resonance, *Trans. Roy. Soc. (London)*, 1870, **A161**, 77–118.
9. RITZ, W., Über eine neue Methode zur Lösung gewissen Variations—Probleme der mathematischen Physik, *J. Reine Angew. Math.*, 1909, **135**, 1–61.
10. GALERKIN, B. G., Series solution of some problems of elastic equilibrium of rods and plates (Russian), *Vestn. Inzh. Tech.*, 1915, **19**, 897–908.
11. NORRIE, D. and DE VRIES, G., *Finite Element Bibliography*, Plenum Press, London, 1976.
12. TREFFTZ, E., Gegenstück zom Ritz'schen Verfahren, *Proc. 2nd Int. Congress Applied Mechanics*, Zurich, 1926.
13. MASSONNET, C. E., Numerical use of integral procedures, Chapter 10 in *Stress Analysis* (Zienkiewicz, O. C. and Holister, G. S. eds.) Wiley, New York, 1965.
14. RIZZO, F. J., An integral equation approach to boundary value problems of classical elastostatics, *Q. J. Appl. Math.*, 1967, **25**, 83–95.
15. CRUSE, T. A., Application of the boundary-integral equation method to 3D stress analysis, *J. Comp. Struct.*, 1973, **3**, 509–27.
16. BUTTERFIELD, R. and BANNERJEE, P. K., The elastic analysis of compressible piles and pile groups, *Geotechnique*, 1971, **21** (No. 1), 43–60.
17. ZIENKIEWICZ, O. C., The finite element and boundary solution methods as general procedures of approximation to field problems. *Proc. World Congress in Finite Element Method in Structural Mechanics*, Bournemouth 1975.
18. ZIENKIEWICZ, O. C., KELLY, D. W. and BETTESS, P., The coupling of the finite element method and boundary solution procedures, *Int. J. Num. Meth. Eng.*, 1977, **11**, 355–76.
19. CRANDALL, S. H., *Engineering Analysis*, McGraw-Hill, London and New York, 1956.
20. FINLAYSON, B. A., *The Method of Weighted Residuals and Variational Principles*, Academic Press, London and New York, 1972.
21. QUINLAN, P. M., The edge-function method in elasto-statics, in *Studies in Numerical Analysis*, Academic Press, London and New York, 1974.
22. VON KARMAN, T., *Calculation of pressure distribution on airship hulls*, NACA TM 574, 1930.
23. BANNERJEE, P. K. and BUTTERFIELD, R., Boundary element methods in geomechanics, *Finite Elements in Geomechanics* (Gudehus, G., ed.), Wiley, New York, 1977, pp. 529–70.

24. ZIENKIEWICZ, O. C., KELLY, D. W. and BETTESS, P., Marriage à la mode, finite elements and boundary integrals, in *Proc. Conf. Innovative Numerical Analysis in Engineering Science*, *CETIM*, Paris, 1977.

25. KELLY, D. W., MUSTOE, G. G. W. and ZIENKIEWICZ, O. C., A hierarchical order for basis functions based on satisfaction of the government equations, in *First Int. Conf. on Recent Advances in Boundary Element Method*, Southampton, July, 1978.

26. LACHAT, J. C. and WATSON, J. O., Effective numerical treatment of boundary integral equations: A formulation for three-dimensional elastostatics. *Int. J. Num. Meth. Eng.*, 1976, **10**, 991–1006.

27. VARGA, R. S., *Matrix Iterative Analysis*, Prentice-Hall, 1962.

28. GRIFFIN, D. S. and KELLOG, R. B., A numerical solution of axially symmetrical and plane elasticity problems. *Int. J. Solids and Structures*, 1967, **3**, 781–94.

29.(a) WILKINS, M. L., Calculation of elasto-plastic flow, Chapter 6 in *Methods in Computational Physics* 3 (Alder, B. *et al.*, eds.) Academic Press, New York, 1964.

  (b) WILKINS, M. L., Finite Difference Scheme for Calculating Problems in Two Space Dimensions and Time, *J. of Computational Physics*, 1970, **5**, 406–14.

30. IRONS, B. M. and RAZZAQUE, A., Experience with the patch test for convergence of finite elements method, pp. 557–87, in *Mathematical Foundations of the Finite Element Method* (Aziz, A. R., ed.) Academic Press, London and New York, 1972.

31. FRAEIJS DE VEUBEKE, B., Variational principles and the patch test, *Int. J. Num. Meth. Eng.*, 1974, **8**, 783–801.

32. AZIZ, A. K. (ed.), *The Mathematical Foundations of the Finite Element Method*, Academic Press, London and New York, 1972.

33. MELOSH, R. J. and MARCAL, P. V., An Energy Basis for Mesh Refinement in Structural Continua, *Int. J. Num. Meth. Eng.*, 1977, **11**, pp. 1083–91.

34.(a) BABUSKA, I., The Self-Adaptive Approach in the Finite Element Method, in *The Mathematics of Finite Elements and Applications* (Whiteman. J. R., ed.), Academic Press, London and New York, 1976.

  (b) PEANO, A. G., SZABO, B. A. and MEHTA, A., Self-adaptive finite elements in fracture mechanics. In preparation.

35. PIAN, T. H. H., Hybrid models, in *Numerical and Computer Methods in Applied Mechanics* (Fenves, S. J. *et al.*, eds.), Academic Press, London and New York, 1971.

36. RASHID, Y. R. and ROCKENHAUSER, W., Pressure vessel analysis by finite element techniques, *Proc. Conf. Prestressed Concrete Pressure Vessels*, Inst. Civ. Eng. London, 1968.

37. BELYTSCHKO, T., CHIAPETTA, R. L. and BARTEL, H. D., Efficient large scale non-linear transient analysis by finite elements, *Int. J. Num. Meth. Eng.*, 1976, **10**, 579–96.

38. SHANTARAM, D., OWEN, D. R. J. and ZIENKIEWICZ, O. C., Dynamic transient behaviour of two and three dimensional structures including plasticity, large deformation and fluid interaction, *Int. J. Earthquake Eng. Struct. Dynam.*, 1976, **4**, 561–78.

39. OTTER, J. R. H., CASSEL, A. L. and HOBBS, R. E., Dynamic relaxation, *Proc. Inst. Civ. Eng.*, 1966, **35**, 633–56.

40. OTTER, J. R. H., Dynamic relaxation compared with other iterative treatments of partial differential equations, *Mech. Eng. Designs*, 1966, **3**, 183–5.
41. BREW, J. S. and BROTTON, D. M., Non-Linear Structural analysis by dynamic relaxation, *Int. J. Num. Meth. Eng.*, 1971, **3**, 463–83.
42. ZIENKIEWICZ, O. C., and PHILLIPS, D. V., An automatic mesh generation scheme for plane and curved surfaces by isoparametric coordinates, *Int. J. Num. Meth. Eng.*, 1971, **3**, 519–28.
43. KAMEL, H. A., and EISENSTEIN, H. K., Automatic mesh generation in two and three dimensional interconnected domains, pp. 455–76, in *High Speed Computing of Elastic Structures*, IUTAM, Univ. Liège, 1971.
44. FRANCAVILLA, A. and ZIENKIEWICZ, O. C., Presentation of three-dimensional stress for dam analysis—computer graphics, *Numerical Analysis of Dams*, University College, Swansea, 1975, pp. 387–94.
45. MOTE, C. D. JR., Global–local finite element, *Int. J. Num. Meth. Eng.*, 1971, **3**, 565–74.
46. BETTESS, P., Infinite elements, *Int. J. Num. Meth. Eng.*, 1977, **11**, 53–64.

Chapter 2

THE METHOD OF CAUSTICS APPLIED TO ELASTICITY PROBLEMS

P. S. THEOCARIS

The National Technical University, Athens, Greece

SUMMARY

The optical method of caustics *was applied to the solution of the singular-stress fields related to a large number of engineering problems. According to this method, the singular region of the problem considered is transformed by means of purely geometric-optics concepts to a highly illuminated curve, received on a reference screen at some distance from the plate. While measurements are difficult and inaccurate to make in the singular region of the problem by conventional methods of experimental stress analysis, the caustic formed on the screen enables the accurate evaluation of its geometrical elements, from which the characteristic parameters of the considered stress field can be easily determined. The method was applied to the following problems of mechanics: (i) to cracked plates loaded in the elastic and plastic regions for the determination of the stress intensity at the crack tips, the length of the plastic zone and the crack-opening displacements, (ii) to cracked shells for the determination of the stress intensity factors at the crack tips, (iii) to dynamic crack-propagation problems for the determination of the stress intensity factors and the strain energy release rates, (iv) to composite materials for the determination of the order of stress singularity and the stress intensity at the region where the constituent bodies coalesce, (v) to contact problems for the determination of the stress distribution in the contacting surface, and (vi) to the solution of laterally loaded flexed plates.*

27

INTRODUCTION

Recently, an optical method has been developed at the Laboratory for Testing Materials of the National Technical University of Athens for the determination of the characteristic parameters of singular elastic or plastic stress fields. In these cases the corresponding stress parameters at the particular points of the problem considered are governed by singularities, which render difficult the solution of the problem by conventional stress analysis methods. The difficulty arises from the fact that the highly strained region near the singular point is infinitesimal and the information obtained by the well-known methods of experimental stress analysis is rather vague. Indeed, in the optical patterns derived from the typical methods of photomechanics, like photoelasticity, interferometry, holography, etc., the region near the singular point appears as a black dot and no substantial information can be gathered from it. In such cases, researchers have developed special evaluation techniques of the corresponding optical patterns, based on the use of suitable extrapolation laws. However, such analyses yield crude results and they cannot be accepted for an accurate evaluation of the stress parameters.

In the present chapter, a new method, called *the method of caustics*, is used for the determination of the characteristic parameters of a large number of singular stress fields of special interest in practical applications. According to the method of caustics, the singular stress field, governed usually by stress singularities, is transformed by means of purely geometrical optics concepts to a dark region bounded by a highly illuminated curve, the *caustic*. The caustic is formed by letting a light bundle to impinge on the specimen and considering the reflected-light rays from either the front or the rear faces of the specimen or those traversing the specimen. The scattered reflections from the anomalous region of the stress field, due to the steep variation of the stress parameters at this region, are concentrated along a singular surface in space. When this surface is cut by a reference plane, a highly illuminated curve, called *caustic*, is created.

The caustic, engendered by the reflected or the transmitted light rays from the anomalous region of the stress field, contains all the necessary information for the determination of the characteristic parameters of the corresponding stress field. This can be achieved by interrelating the characteristic anomalous stress-field parameters with the geometrical elements of the caustic. These latter quantities can be measured with a high accuracy, thus enabling the determination of the basic parameters of the singular stress field.

The method of caustics is used for the solution of several elasticity problems presenting particular interest in engineering applications. In most of these problems the method of caustics will be used to study the singular stress fields near crack tips, wedge apices, etc. Nevertheless, some cases where no stress field singularities are present will be also studied. In all cases, the experimental and theoretical results are in good agreement. Finally, it should be mentioned that the contents of this chapter constitute a review of the author's work on the method of caustics. Several more results on the method of caustics can be found in the references given at the end of the chapter.

THE OPTICAL METHOD OF CAUSTICS

The General Theory of the Method of Caustics[1]

Let us consider a surface with equation:

$$z = f(x, y) \tag{1}$$

and a parallel-light beam illuminating the surface (Fig. 1). The rays of this beam are reflected on the surface and deviate from parallelism. If a reference screen is placed parallel to the plane Oxy, to which the above surface is referred, and at some distance z_0 from it, then the deviation $\mathbf{w}$ on

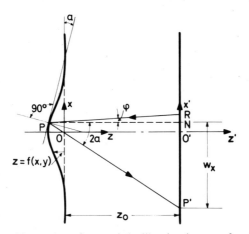

Fig. 1. Geometry of formation of a caustic by illuminating a surface by a parallel or divergent light beam.

the screen of the reflected ray from a point $P(x, y)$ of the surface is given, according to Snell's law of reflection, by (Fig. 1):

$$\mathbf{w} = w_x\mathbf{i} + w_y\mathbf{j} \tag{2}$$

with:

$$w_x = (z - z_0)\tan 2\alpha \qquad w_y = (z - z_0)\tan 2\beta \tag{3}$$

$$\tan \alpha = \frac{\partial f(x, y)}{\partial x} \qquad \tan \beta = \frac{\partial f(x, y)}{\partial y} \tag{4}$$

where $\mathbf{i}$ and $\mathbf{j}$ are Cartesian unit vectors, referred to the projection $O'x'y'$ of the frame Oxy on the screen.

If we refer the vector $\mathbf{w}$ on the origin O' of the system $O'x'y'$, we obtain that the image on the screen of any point $P(x, y)$ of the surface is given by:

$$\mathbf{W} = W_x\mathbf{i} + W_y\mathbf{j} \tag{5}$$

with:

$$W_x = x + [f(x, y) - z_0]\frac{2\partial f(x, y)/\partial x}{1 - (\partial f(x, y)/\partial x)^2}$$
$$W_y = y + [f(x, y) - z_0]\frac{2\partial f(x, y)/\partial y}{1 - (\partial f(x, y)/\partial y)^2} \tag{6}$$

Relations (6) map each point $P(x, y)$ of the surface $z = f(x, y)$ to a point $P'(W_x, W_y)$ on the screen (Fig. 2). The necessary and sufficient condition that the points P' on the screen belong to a curve is the zeroing of the

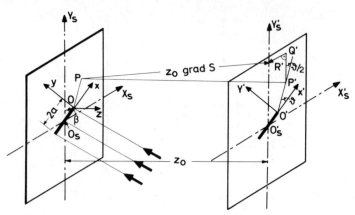

FIG. 2.   Geometry of cracked plate and relative position of specimen and viewing screen.

Jacobian determinant of the transformation defined by relations (6). This determinant is expressed as:

$$J = \frac{\partial(W_x, W_y)}{\partial(x, y)} = \begin{vmatrix} \dfrac{\partial W_x}{\partial x} & \dfrac{\partial W_x}{\partial y} \\[2mm] \dfrac{\partial W_y}{\partial x} & \dfrac{\partial W_y}{\partial y} \end{vmatrix} = 0 \qquad (7)$$

Relation (7) defines a curve on the surface $z = f(x, y)$, called the *initial curve* and the system of eqns. (6) and (7) defines on the screen its corresponding *caustic*.

If the elevations and the slopes of the surface $z = f(x, y)$ are small, the squares of the derivatives of the function $f(x, y)$ can be neglected, when compared to unity, and if, further, the elevation $z = f(x, y)$ is negligible relative to z_0, we obtain from relations (6) that:

$$W_x = x - 2z_0 \frac{\partial f(x, y)}{\partial x}$$

$$W_y = y - 2z_0 \frac{\partial f(x, y)}{\partial y} \qquad (8)$$

If the incident light beam subtends an angle φ with the normal to the screen at a point $P(x, y)$, then we must replace 2α and 2β by $(2\alpha + \varphi)$ and $(2\beta + \varphi)$ respectively in the expressions for the components of the deviation vector **w** (see Fig. 1).

In the special case of a point light source at a distance z_i from the surface, we obtain in the case of small slope changes of the surface the following expression for the components $W_{x,y}$:

$$W_{x,y} = x, y + [f(x, y) - z_0] \tan(2\alpha, \beta + \varphi) \qquad (9)$$

or:

$$W_{x,y} \simeq x, y - z_0(2 \tan \alpha, \beta + \tan \varphi)$$

which can be written as:

$$W_{x,y} = \lambda_m x, y + w_{x,y} \qquad (10)$$

where λ_m is the magnification factor, defined by:

$$\lambda_m = \frac{z_0 + z_i}{z_i} \qquad (11)$$

if the point-light source illuminating the surface lies on the z-axis.

Application to Plane Elasticity Problems[2-7]

In plane elasticity problems the above defined surface $z = f(x, y)$ represents the variation of the half of the thickness t of the two-dimensional specimen, due to loading. The variation Δt of the thickness of a plane specimen is given, according to Poisson's ratio effect, by:

$$\Delta t = -\frac{vt}{2E}(\sigma_1 + \sigma_2) \qquad (12)$$

where v and E represent Poisson's ratio and the elastic modulus of the material of the specimen respectively and σ_1 and σ_2 are the two principal stresses.

For the case of a plane specimen loaded in the elastic range, the variation of its thickness is small when compared to the thickness itself, so then eqn. (10) may be used, without any loss of accuracy, instead of relations (9). Thus, it is obtained for the vector $\mathbf{W}(W_x, W_y)$:

$$\mathbf{W} = \lambda_m \mathbf{r} + z_0 \frac{vt}{E} \operatorname{grad}(\sigma_1 + \sigma_2) \qquad (13)$$

Equation (13) gives the deviation vector of a light ray reflected from the front face of a plane specimen. In the case when a light beam is allowed either to pass through the specimen, or to be reflected from the rear face of the specimen, then the variation of the optical path of the corresponding light ray due to loading must also be taken into account.

The variation of the optical path of a light ray, either reflected from the rear face of the specimen, Δs_r, or traversing the specimen, Δs_t, as was already shown (2), is given by:

$$\Delta s_{r1,2} = 2tc_r[(\sigma_1 + \sigma_2) \pm \xi_r(\sigma_1 - \sigma_2)]$$
$$\Delta s_{t1,2} = tc_t[(\sigma_1 + \sigma_2) \pm \xi_t(\sigma_1 - \sigma_2)] \qquad (14)$$

where the indices 1 and 2 refer to the two principal-stress directions and c_r, c_t, ξ_r, and ξ_t are stress optical constants, characteristic of the material of the considered medium (2).

By taking into account relations (14) in combination with relation (13) we can express the deviation vector $\mathbf{W}$ of a light ray reflected either from the front, or the rear face of the specimen, or transmitted through the specimen, for the case of an optically inert material ($\xi = 0$) by the following universal relation:

$$\mathbf{W} = \lambda_m \mathbf{r} + C \operatorname{grad}(\sigma_1 + \sigma_2) \qquad (15)$$

where the global constant C takes the following values for the case of the light rays reflected from the front (C_f), or the rear face of the specimen (C_r), or transmitted through the specimen (C_t) respectively:

$$C_f = z_0 t c_f \qquad c_f = v/E \tag{16}$$

$$C_r = -2z_0 t c_r \tag{17}$$

$$C_t = -z_0 t c_t \tag{18}$$

In the case of the isotropic elastic materials when the sum $(\sigma_1 + \sigma_2)$ of the principal stresses is expressed by an analytic function $\Phi(z)$:

$$\sigma_1 + \sigma_2 = 4\,\mathrm{Re}\,\Phi(z) \tag{19}$$

where the symbol Re denotes the real part of a complex quantity, relations (7) and (15), yielding the caustic and its initial curve respectively, take the following simplified forms:

$$\mathbf{W} = \lambda_m z + 4C\left(\overline{\frac{d\Phi(z)}{dz}}\right) \tag{20}$$

$$\left|\frac{4C}{\lambda_m}\frac{d^2\Phi(z)}{dz^2}\right| = 1 \tag{21}$$

CRACKS IN PLANE ISOTROPIC ELASTIC MEDIA[8-16]

For the case of a crack in an infinite plate subtending an angle β with the axis of loading, the corresponding complex stress-function $\Phi(z)$ is expressed in the close vicinity of the crack tip by:

$$\Phi(z) = \frac{K^*}{2(2\pi z)^{1/2}} \tag{22}$$

where K^* is the complex stress intensity factor, given by:

$$K^* = K_I - iK_{II} = |K^*|\exp(-i\omega) \qquad \frac{K_{II}}{K_I} = \tan\omega = \mu \tag{23}$$

with K_I and K_{II} representing the *opening mode* and *sliding mode stress intensity factors* respectively.

Introducing eqn. (22) into relations (20) and (21), we obtain for the initial curve and the corresponding caustic respectively:

$$r_0 = \left|\frac{3CK^*}{2\lambda_m(2\pi)^{1/2}}\right|^{2/5} \tag{24}$$

$$\mathbf{W} = \lambda_m r_0 (\exp(i\vartheta) + \tfrac{2}{3}\exp[(3\vartheta/2 + \omega)i]) \tag{25}$$

where λ_m is the magnification factor of the optical arrangement and ϑ is the polar angle of a system of polar coordinates with its origin at the crack tip.

Equation (24) indicates that the initial curve is a *circle* of radius r_0, while the corresponding caustic is a *generalised epicycloid*.

Equation (25), when expressed in parametric form, becomes:

$$X = \lambda_m r_0 \left[\cos \vartheta \pm \frac{2}{3}(1 + \mu^2)^{-1/2} \cos\left(\frac{3\vartheta}{2}\right) \mp \frac{2}{3}\mu(1 + \mu^2)^{-1/2} \sin\left(\frac{3\vartheta}{2}\right) \right]$$

$$Y = \lambda_m r_0 \left[\sin \vartheta \pm \frac{2}{3}(1 + \mu^2)^{-1/2} \sin\left(\frac{3\vartheta}{2}\right) \pm \frac{2}{3}\mu(1 + \mu^2)^{-1/2} \cos\left(\frac{3\vartheta}{2}\right) \right]$$

$$(26)$$

with:

$$r_0 = \left| \frac{3CK^*}{2\lambda_m(2\pi)^{1/2}} \right|^{2/5} \qquad \mu = \tan \omega \qquad (27)$$

Figure 2 presents the formation of caustic by illuminating a cracked plate in tension.

If we refer the epicycloid to a new coordinate system with the same origin, but angularly displaced by an angle (-2ω), we obtain for the parametric equations of the epicycloid the relations:

$$X = \lambda_m r_0 \left(\cos \varphi \pm \frac{2}{3}\cos\frac{3\varphi}{2} \right)$$

$$Y = \lambda_m r_0 \left(\sin \varphi \pm \frac{2}{3}\sin\frac{3\varphi}{2} \right)$$

$$(28)$$

with $\varphi = \vartheta + 2\omega$.

From these two relations it is concluded that the epicycloid is a symmetrical curve about an axis subtending an angle equal to (-2ω) to the crack line. Furthermore, if the crack-axis in a plate under simple tension subtends an angle β with the axis of loading σ, then K_I and K_{II} are given by:

$$K_I = \sigma(\pi\alpha)^{1/2} \sin^2 \beta \qquad K_{II} = \sigma(\pi\alpha)^{1/2} \sin\beta\cos\beta \qquad (29)$$

so that we obtain from the second of relations (23):

$$\frac{K_{II}}{K_I} = \cot \beta = \tan \omega \qquad (30)$$

Thus, angle ω must be equal to the inclination of the axis of the crack with the transverse axis of the plate. Therefore, for plates in simple tension

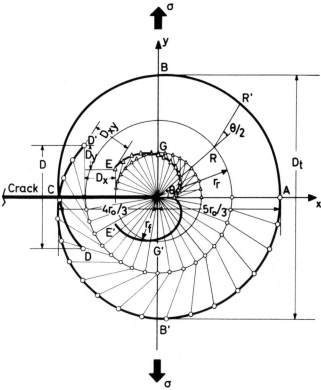

FIG. 3.   Shape of the principal epicycloid and geometry of its formation for
$$c_f/c_r = -0.25.$$

the axis of symmetry of the caustics is symmetric to the crack-axis with
respect to the axis of loading of the plate.

From eqns. (29) we obtain that the polar radius ρ of the caustic presents
an extremum for $\varphi = 0$ where it is valid that:

$$\rho_{max} = \tfrac{5}{3}r_0 \qquad \text{or} \qquad \rho_{min} = \tfrac{1}{3}r_0 \qquad (31)$$

Furthermore, it can be derived that the diameter of the caustic along its axis
of symmetry is equal to $D_1 = 3r_0$, while the transverse diameter of the
caustic is $D_t = 3.17r_0$.

Figure 3 presents the caustic and its geometrical construction for a crack
subtending an angle $\beta = 90°$ to the axis of the applied stress σ at infinity for
the case when $C_f/C_r = -0.25$. The external part of the caustic (curve
D'CB'ABCD), the geometrical construction of which is shown in this

figure, is formed by reflections from the rear face of the plate and corresponds to $C_r = -2z_r t c_r$. This part stops the reflections from the lips of the crack (points D and D′ in Fig. 3). The internal part of the caustic is formed from the front face (curve E′G′GE) of the plate with $C_f = z_0 C_f . t$. While for $C_f = C_r$ the two parts of the caustic have two common points, when $C_f \neq C_r$, a discontinuity between these two parts exists. In the case of

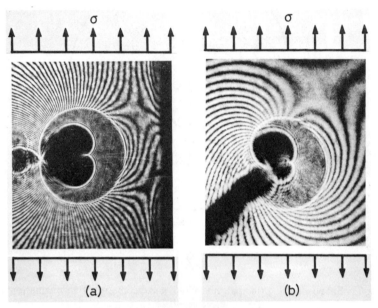

FIG. 4. Interference patterns of the constrained zones around cracks for $\beta = 90$ deg. (a) and 45 deg. (b).

plexiglas plates, the ratio of the radii of the corresponding initial curves of the two parts of the reflected caustics is equal to $r_r/r_f = 1·48$.

In Fig. 4 the experimentally obtained caustics from two cracked plexiglas specimens (a) and (b) with $\beta = 90$ deg. and 45 deg. respectively are shown. It can be observed from Fig. 4(b) that the axis of symmetry of the caustic is inclined at an angle $-45°$ with the vertical axis of the specimen, which is in accordance with the previously defined result that the axis of symmetry of the caustic is symmetric to the crack-axis with respect to the axis of loading of the plate. Furthermore, Fig. 5 presents the variation of the difference of longitudinal diameters of the generalised epicycloids $(D_1^{max} - D_1^{min})$, normalised to the maximum longitudinal diameter D_1^{max},

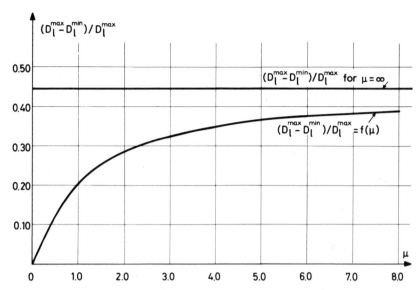

FIG. 5. Difference of longitudinal diameters of generalised epicycloids $(D_I^{max} - D_I^{min})$, normalised to the maximum longitudinal diameter D_I^{max} versus the ratio $\mu = K_{II}/K_I$.

versus the ratio $\mu = K_{II}/K_I$. By measuring this quantity from the experimentally obtained caustics, the value of μ can be determined. Then, the values of K_I and K_{II} can be found from the following relations, which can be deduced from eqns. (26) and (27):

$$K_I = \frac{\sqrt{2\pi} r_0^{5/2}}{z_0 t c \sqrt{1 + \mu^2}} \qquad K_{II} = \mu K_I \qquad (32)$$

Figure 6 shows a representative picture of the formation of the caustic on the reference screen at some distance from the plate for the case of an edge-cracked tension plate. The caustic is formed from a very small initial curve S on the surface of the specimen. This initial curve, projected on the screen, forms the caustic, which is highly enlarged. The enlargement of the caustic is due to the steep-thickness and refractive-index variations of the specimen at the vicinity of the crack tip, which scatter the reflected-light rays, as well as to the magnification of the optical arrangement.

The incident-light beam on the surface of the specimen may be parallel, convergent or divergent. As it can be easily concluded from relations (24) and (25), when the incident-light beam is convergent or divergent, the initial

38 P. S. THEOCARIS

curve becomes smaller than for the case of a parallel-light beam, while the
caustic becomes larger, because an additional enlargement exists due to the
divergence of the light rays. Thus, when the incident light beam is non-
parallel the initial curve, becoming very small, gathers information from
the close vicinity of the crack tip. This is of particular importance and
proves the superiority and the powerfulness of the method of caustics for

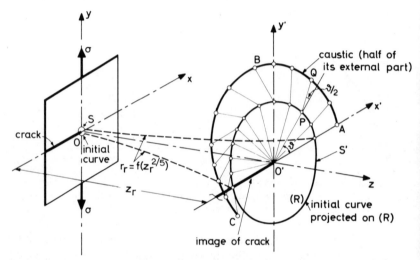

Fig. 6.   Schematic of formation of caustic in a distance z_r from the cracked plate.

the solution of problems with cracks over the conventional methods of
experimental stress analysis.

The above developed theory for the study of a single inclined crack in an
elastic plate subjected to a remote uniaxial tension can similarly be
extended to solve problems with complicated geometric crack configurations and loading conditions. A large variety of such problems has
been solved by the method of caustics.[17-21] Among them the problems of
an array of either equal, or unequal cracks and the problem of a bifurcated
crack have been thoroughly studied. Crack problems in time-variable
viscoelastic stress fields have also been studied by the method of
caustics.[22-23]

The crack opening displacement that is the distance between the crack
lips due to loading, which is considered also as a characteristic parameter of
fracture mechanics, can be determined by the method of caustics.[24] Indeed,
because of the existing distance between the lips of the crack, there is always

a gap between the external and the internal branches of each caustic. Furthermore, since the external part of the caustic is made by reflection from the rear face (for tension specimens and diverging light beams impinging the plate) and the internal part is made by reflections from its front face, the optical constants C_r and C_f are different and therefore each branch of the caustic presents also a shift relative to the other. In the case when the specimen is transparent and both caustics, formed by reflection from the front and rear faces of the specimen, are obtained, the existing gaps between these two caustics provide additional information for measuring the crack opening displacement. For a more detailed analysis of the application of the method of caustics to the measurement of the crack opening displacements we refer to reference 24. In this reference a series of experimental results is also presented.

AN OPTICAL STRESS-ROSETTE BASED ON CAUSTICS[25]

Caustics are not devoted exclusively to the study of the stress field at the vicinity of singularities. They can also be applied to the study of the stress distributions near any stress concentration, since these concentrations create lateral deformation of the specimen due to Poisson's effect. In the sequence, we develop a method for studying stress concentrations in elastic fields. This method is called the *optical stress rosette*. According to this method, the principal directions and the difference of the principal stresses can be directly determined, i.e. this rosette constitutes a direct substitute for the old method of photoelasticity. The rosette is based on the caustics formed by a small circular hole drilled in a two-dimensional elastic stress field or any axisymmetric protuberance on the surface of the specimen. Then, it can be proved that the axes of symmetry of the caustics coincide with the principal stress directions, whereas the difference of the principal stresses is connected to the transverse diameter D of the caustic formed, by the relation:

$$\sigma_1 - \sigma_2 = \frac{3^3}{2^{10}} \frac{D^4}{C R^2} \tag{33}$$

where R is the radius of the hole or protuberance.

In Fig. 7 the experimentally obtained caustics formed by the reflected light rays from the two faces of a drilled plexiglas specimen are shown. These caustics are in accordance with the expected corresponding theoretical forms.

FIG. 7. Interferogram and caustics formed by reflected rays at the front and rear faces of the plate.

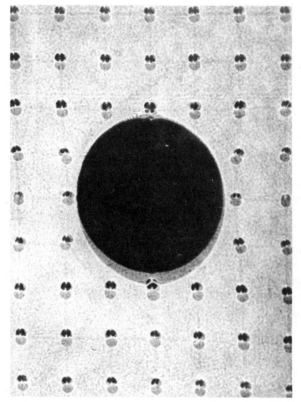

FIG. 8. Reflected image of the stress field in a plate containing a central symmetric hole and subjected to tension as it is depicted by an array of stress rosettes, (a) reflected light, (b) traversing light.

The method was applied for analysing the stress field around a circular hole in a tension specimen. Figure 8 gives the optical pattern obtained by illuminating the tension specimen with a large circular perforation, in which a number of small circular holes has been drilled at the knots of a square network. The caustics obtained around the holes were used to determine the principal directions and the difference of the principal stresses. The results obtained are in agreement with the theory.

The importance of this application lies in the fact that at the rim of the perforation there is no stress singularity. Only the stress concentration around the hole creates a thickness variation which is the reason for the formation of the double kidney-like caustic. It is worth while mentioning here that either a distinct initial curve exists, which is a circle concentric to the perforation at the interior of the plate, or the boundary of the hole plays the role of the initial curve. This depends on the amount of load applied to the plate and on its geometry. However, in the second case, where a boundary of the deformed plate substitutes the initial curve, the caustic created is not a *real caustic*, defining the part of the field where reflections from both sides of the corresponding initial curve are exclusively concentrated in this part, while the other part of the field remains completely dark. In this second case the caustic formed is simply due to the fact that the image of the deformed boundary creates on the screen a curve of concentration of reflected light rays from the vicinity of this boundary. Since there is no material beyond the boundary of the plate, this limiting curve, which is called a *pseudocaustic*, has analogous properties with those of the real caustics.

OTHER APPLICATIONS TO CRACK PROBLEMS

In this section we will mention the application of the method of caustics to three more interesting crack problems, that is: cracked elastic–plastic media; cracked cylindrical shells; as well as dynamically propagated cracks.

Cracks in Elastic–Plastic Stress Fields[26–29]
The method of caustics can also be applied to the study of crack problems in elastic–plastic materials. In this case some simple model should be used to represent the stress field near a crack tip. Such a model is the well-known Dugdale–Barenblatt model, valid for elastic–perfectly plastic materials, or its modified form shown in Fig. 9. In this figure it is assumed that the whole specimen is in an elastic stress state except along two short straight

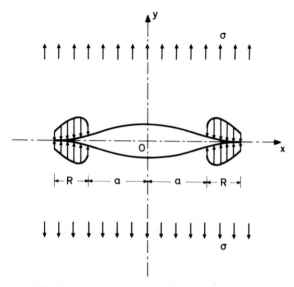

FIG. 9.   The modified form of the Dugdale–Barenblatt model for an internal crack
in the elastic–perfectly plastic material.

segments adjacent to the crack tips where the stress distribution takes values
exceeding the elastic limit. The caustics formed under these conditions have
been studied in detail in references 26–29 and can provide useful
information concerning the length of the plastic zones under the foregoing
assumptions. Of course, the equations of the caustics and their initial curves
evidently become sufficiently complicated in the present case and can be
found in references 26–29.

In Fig. 10 two theoretical forms of caustics in elastic–perfectly plastic
specimens under the assumption of validity of the simple Dugdale–
Barenblatt model are presented. In these caustics the ratio of the stress σ at
the elastic limit to the loading σ_0 at infinity and normally to the crack was
assumed equal to $\sigma/\sigma_0 = 0\cdot20$ (Fig. 10a) and $0\cdot30$ (Fig. 10b). The forms of
the caustics shown in Fig. 10 were seen to be similar to the corresponding
experimentally obtained caustics.

The application of the method of caustics for the study of crack problems
in the elastic–plastic region of loading by using the Dugdale–Barenblatt
model is justified because in this model it is assumed that the plastic enclave
is extended along the crack-axis in narrow strips forming a protrusion
ahead of the crack. Thus, all the area surrounding the crack-tip, except this
narrow protrusion, is assumed to be in the elastic region of loading and

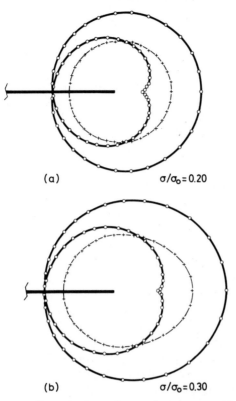

(a) $\sigma/\sigma_0 = 0.20$

(b) $\sigma/\sigma_0 = 0.30$

FIG. 10.   Four types of caustics formed around a transverse edge crack for values
of ratio σ/σ_0 0·20 (a) and 0·30 (b) for the simple D–B model.

therefore the elastic theory, on which the formation of the caustic is based,
is still valid except for the narrow zone at the crack-axis where plasticity is
developed.

However, only a few materials present such types of plastic zones under
special modes of loading. This was immediately recognised by Dugdale who
always found a discrepancy between the theoretical predictions of his model
and the experimental results. In order to improve the results of the model by
taking into consideration the influence of strain-hardening of the material
at the plastic area of loading, the triaxiality of the stress field in the vicinity
of the crack and the fact that the maximum tensile stress normal to the
crack-axis in the plastic enclave appears inside the plastic zone, the modified
D–B model was introduced by Rosenfield, Dai and Hahn[31] and applied to
caustics as in references 26 to 29.

According to this modified D–B model the distribution of the normal to the crack-axis stress at the interior of the plastic enclave is considered as a step function varying between the yield limit σ_0 in pure tension and the maximum tensile stress encountered in the interior of the enclave. Experimental evidence indicated that in all cases this $\sigma_{0\ max}$ cannot exceed 10 to 40 percent of σ_0. The number of steps in the step function depends on the approximation imposed on the solution. For each step the simple Dugdale–Barenblatt model was applied with σ_0 as the corresponding stress to this particular step according to the distribution profile assumed in the plastic enclave.

In this way we succeeded in making the model more adaptable to real cases of plastic deformation of metals. Six different stress-profiles were selected which resemble the normal stress distributions in the plastic enclaves for different materials and different degrees of strain-hardening. The corresponding caustics for these profiles were traced in a computer for a variety of values for the ratio $\sigma_{0\ max}/\sigma_0$.

Each time, when a new test is undertaken with a cracked metallic plate, the corresponding caustic for each step of loading must be compared to the series of caustics prepared in advance and corresponding to typical stress-profiles. The coincidence or close similarity between the experimental caustic and the caustic computed for a particular stress profile is the basis in selecting this particular profile as the one corresponding to the case under study. Thus, the caustics obtained during the test provide a means for selecting the closer to reality solution of the problem.

Again, since for the definition of the caustic the validity of the theory of elasticity is assumed, as for the case of the simple model, it is necessary to satisfy the condition that the initial curve of the caustic lies for its larger part in the elastic region as it does with the simple D–B model. However, it is natural to accept that, since the normal to the crack-axis σ_y-stress is larger than σ_0 along the extension of the crack-x-axis for the strain-hardening material the protruding plastic zone along the x-axis must be thicker than the zone for the simple model. Nevertheless, it was experimentally found[30] that the variation of the σ_y-stress in a direction normal to the crack-axis is rather abrupt and therefore the thickness of the plastic zone is not increased significantly as compared to the case of an elastic–perfectly plastic material. Therefore, it may be assumed that the theory of caustics is still valid for the strain-hardening materials if it is valid for the perfectly plastic ones, provided that care was taken to increase the size of the initial curve surrounding the crack tip. This can be done by using an optical arrangement with $\lambda_m = 1$ (parallel-light beam, see relation (24) of the text) and to make the distance z_0 as large as possible, in order to increase the value of the global constant C.

A study concerning the degree of accuracy of the modified D–B model for typical cases of cracked plates made of strain-hardening materials is actually undertaken in our own laboratory.

Cracked Cylindrical Shells[32–33]

When a crack is developed in a shell, the method of caustics can be used for the determination of the stress intensity factor. However, in the case of shells the situation is much more complicated, because, in addition to microscopic variations of curvature of the surface surrounding the crack tip, there is also the macroscopic curvature of the shell, which must be taken into account, since it influences the shape and dimensions of the caustic. Thus, the orientation of the crack plays an important role to the corresponding shape of the caustic at the crack tip. We distinguish two extreme cases, i.e. when the crack is either circumferential or axial. The influence of curvature of the shell in both extreme cases is different.

In the following section we examine, as an example, the case of a cylindrical shell panel with a circumferential crack of length $2a$ subjected to an axial extension and illuminated by a convergent light beam. The equations of the caustics formed at the tips of the crack are given for this problem in references 32 and 33 and they will not be repeated here.

Figure 11 presents the caustics obtained by a cylindrical shell made of polycarbonate and subjected to a tension load equal to $\sigma = 93\cdot0\,\text{kp/cm}^2$ with $z_0/R = 12\cdot09$ and $\lambda_m = 4\cdot87$. While Fig. 11(a) corresponds to the light rays reflected from the shell, the caustic of Fig. 11(b) is formed by the light rays traversing the specimen. It can be observed that only the caustics created by the reflected rays are influenced by the curvature, while the caustic formed by the light rays traversing the specimen is the same as the caustic for a flat plate. In Fig. 12 the theoretical form of the caustic corresponding to the case of Fig. 11(a) ($\lambda_m = 4\cdot87$, $z_0/R = 12\cdot09$, $\xi = 0\cdot153$) for $c_f/c_r = -0\cdot50$ is presented. By comparing Figs. 11 and 12, we observe that the theoretical caustics are very close to the corresponding experimentally obtained forms. From the dimensions of the experimentally obtained caustics and on the basis of the corresponding theoretical equations, the values of the stress intensity factors K at the crack tips can be easily obtained.[32–33]

Dynamic Crack Propagation Phenomena[34–36]

The characteristic quantities of dynamically loaded cracked plates, like the

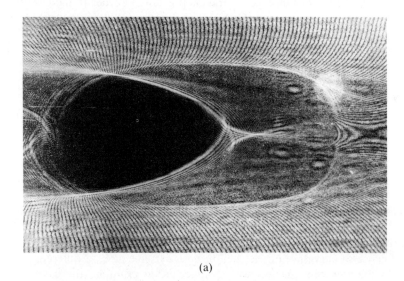

(a)

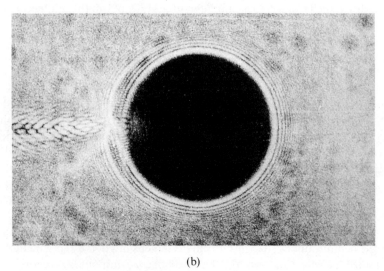

(b)

FIG. 11. Caustics obtained by illuminating a Lexan tension cylindrical shell with $\sigma = 93{\cdot}0\,\mathrm{Kp/cm^2}$, $(z_0/R) = 12{\cdot}09$, $\lambda_m = 4{\cdot}87$. While the caustic of figure (a) is formed by the light rays reflected from the front and rear faces of the shell, the caustic of figure (b) corresponds to the light rays traversing the specimen.

stress intensity factor and the strain energy release rate, can be successfully determined by the method of caustics in combination with a high-speed camera. Two different kinds of dynamic problems were studied, that is either loads high enough to create propagation of crack and rupture, or pulse loads of reduced intensity, which did not succeed to make the crack propagate.

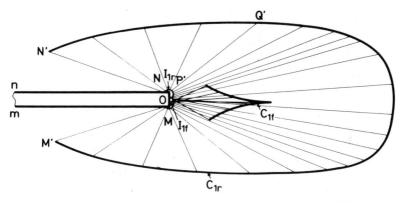

FIG. 12. Caustic and its geometrical construction for the case of Fig. 11(a).

For the case of a load lower than the breaking limit of the specimen the study is concerned with a comparison between the static and the dynamic mechanical and optical properties of the material used, as well as with the stress intensity factors of the specimen configurations considered. The experimental arrangement for the application of a dynamic load to the specimen and the recording of the corresponding optical patterns obtained is presented in reference 34.

The second kind of dynamic problems solved by the method of caustics is related to the study of the mode of fracture of specimens subjected to various types of impact loads. In such problems the mode of propagation of the crack or cracks, the crack velocity, acceleration, deceleration and crack arrest, as well as the variation of the mechanical and/or optical properties and the corresponding dynamic stress intensity factors during this transient phenomenon may be studied experimentally.

Such a case of successive crack propagation and arrest in a tensile specimen made of plexiglas and containing a transverse edge crack subjected to a sequence of compressive and tensile stress pulses is shown in Fig. 13(a). This figure presents a sequence of 11 photographs corresponding to the propagation of a stress pulse creating a strain $\varepsilon_{max} = 3\cdot6 \times 10^3\ \mu$

48 P. S. THEOCARIS

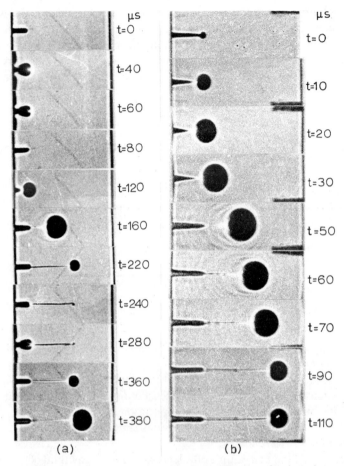

FIG. 13. Optical patterns obtained during the application of an impulse stress to a slotted plexiglas specimen.

strain through this slotted specimen with a slit length $a_0 = 6$ mm. It can be observed from this figure that in the first five photographs the internal part of the caustic presenting a cusp is formed, which means, as has been proven previously[8,11] and also for the optical set-up used in the tests, that these photographs correspond to the propagation of a compressive impulse load through the specimen. During this time-period the crack does not propagate. In the subsequent six photographs the form of the caustic formed at the crack tip is alerted. These circle-shaped caustics reveal that the crack tip stress field is tensile, which corresponds to the tensile part of the reflected impulse stress. It is obvious from this sequence of six photographs that only during this time-interval of the passage of the tensile stress pulse does the crack propagate.

The subsequent series of photographs corresponds again to a compressive stress-pulse created by reflection of the tensile stress-pulse on the other lateral side of the specimen. It can be observed from these photographs that no caustic is formed at the tip of the propagating crack, which is now arrested, but the internal cusp-shaped part of the caustic is again created at the end of the initial slit. Thus, it can be concluded that a real crack in a compressive stress field has no effect on the stress field in the plate as it has on the slit, whose lips lie permanently apart from each other. Contrary to this phenomenon, for the case of the subsequent tensile field both the real crack and the initial notch have the same effect upon the stress field creating both circle-shaped caustics. Finally, the four last pictures of the series correspond to a tensile stress field created by the second passage of the tensile pulse and leading the specimen to fracture.

Since it was proved from a series of preliminary tests, described above, that the compressive impulse stress does not have any influence upon the stress field at the tip of the running crack, only the tensile field was studied further. For this purpose the frequency of the high speed camera was suitably adjusted to detect only the tensile part of the impulsive stress. Figure 13(b) presents 9 from the 24 pictures obtained by the camera during a crack propagation corresponding to the tensile part of the impulsive stress for a slotted plexiglas specimen with an initial slit of length $a_0 = 10$ mm and $\varepsilon_{max} = 3.6 \times 10^3 \mu$ strain.

From the above photographs the variation of the crack length, as well as of the stress intensity factor at the crack tip can be calculated during the time of propagation of the initial slit of the specimens.

For example, by measuring the transverse diameter D_t of the caustic formed at the crack tip, the values of the stress intensity factors K_I can be computed through relation (32). Figure 14 shows the variation of K_I versus the length of the propagating crack for $a_0 = 4$ mm and $\varepsilon_{max} = 7.5 \times 10^3 \mu$ strain, $a_0 = 6$ mm and $\varepsilon_{max} = 5.2 \times 10^3 \mu$ strain and $a_0 = 10$ mm and $\varepsilon_{max} = 3.6 \times 10^3 \mu$ strain respectively. From these curves it can be concluded that there is a critical value of the stress intensity factor for which the crack does not propagate. From this value of the critical stress intensity factor the value of the strain energy release rate G_{Ic} can be calculated, by using the simple Griffith criterion. The calculated value for G_{Ic} was found to be in good agreement with that given by Irwin.[37]

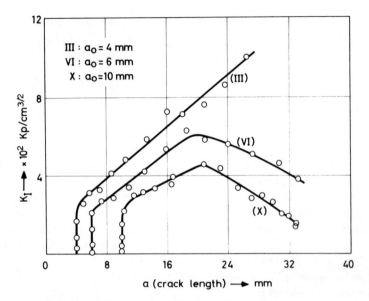

FIG. 14.   Variation of the stress-intensity factor K_I as a function of crack length a.

COMPOSITE MATERIALS[38-40]

The elastic stress field at the point where the various phases of a composite material coalesce are usually governed by stress singularities. The type and the order of these singularities can be studied by the method of caustics. For the most general case the stress function at the vertex of the multiwedge can be expressed by:

$$\Phi_i(z) = K_i z_i^{-p} \qquad z_i = r_i \exp(i\varphi_i) \qquad p = p_1 + ip_2 \qquad (34)$$

where K_i is the stress intensity factor and p is the order of the elastic stress singularity, generally expressed by a complex number.

This expression for the stress function yields the result that the initial curve is a circle with radius:

$$r_i = |4C_i K_i p(p+1) \exp(p_2 \varphi_i)|^{1/(p_1+2)} \qquad (35)$$

whereas the caustic is described by

$$W = r_i \exp(i\varphi_i) \pm 4C\bar{K}_i \bar{p} r_i^{-(\bar{p}+2)} \exp[i(\bar{p}+1)\varphi_i] \qquad (36)$$

From this relation and for the more usual case of real stress singularities the following parametric equations of the caustic are obtained:

$$X_i = r_i \left[\cos \varphi_i \mp \frac{1}{(p + 1)} \cos \left[(p + 1)\varphi_i - \gamma_i \right] \right]$$

$$Y_i = r_i \left[\sin \varphi_i \mp \frac{1}{(p + 1)} \sin \left[(p + 1)\varphi_i - \gamma_i \right] \right] \tag{37}$$

while the corresponding initial curve r_i is expressed by eqn. (35) with $p_2 = 0$.

In these relations the index i is referred to the particular wedge i of the multiwedge, $K_i = |K_i| \exp(i\gamma_i)$. For the special case when the stress intensity factor K_i is real the corresponding equations of the caustic are obtained by putting in the above relations $\gamma_i = 0$.

In Figs. 15 and 16 two typical theoretical and experimental forms of the caustics created at V-notch tips (where the singularity p is real) are shown. From the characteristic dimensions of these caustics and on the basis of the foregoing formulas, the values of the order of singularity p and the stress intensity factor K at the V-notch tip can be easily determined.[38-39]

An interesting application of the method of caustics is the case where a disruption of interface exists between phases in composite materials due to loading. The case of an infinite soft matrix containing a circular hard inclusion was studied in reference 40 when a separation of phases exists of various lengths and orientations.

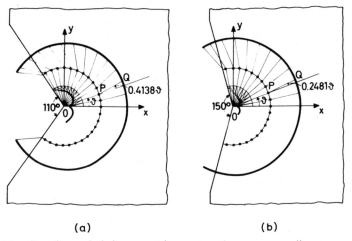

(a)   (b)

FIG. 15. Caustics and their geometric construction corresponding to cases of Fig. 16.

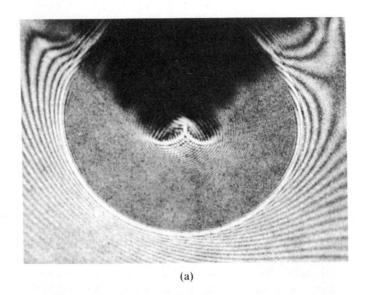

(a)

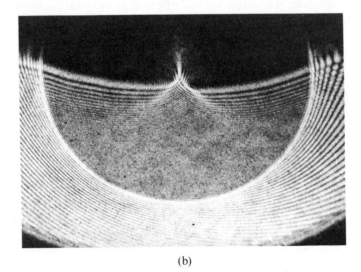

(b)

FIG. 16. Experimentally obtained caustics at the apices of single wedges with
angles $\vartheta = 260°(a)$ and $\vartheta = 210°(b)$.

CONTACT PROBLEMS[41-45]

The problem of the stress distribution at the contacting surface between two bodies has been first considered by the pioneering work of Hertz, and subsequently studied by several authors. Due to the inherent difficulties, the problem has not yet been completely solved. The experimental solution of the problem by the well-known conventional methods meets many difficulties due to the high stress concentration at the contacting area. The method of caustics provides a simple and powerful tool for the solution of the problem. The stress distribution at the contacting surface can be determined by considering the caustic formed by the deformed boundary. These caustics are very sensitive to the form of the stress distribution at the contacting surface and can be easily used for the solution of the inverse problem, that is for the determination of the stress distribution from the corresponding caustics.

Let us consider the case where the boundary of one of the two contacting bodies is straight and let us assume that the stress distribution along this boundary is expressed in a power series form, as follows:

$$P(t) = \sum_{k=0}^{m} S_k t^k \tag{38}$$

Then, the corresponding stress function is given by:

$$\Phi(z) = \frac{1}{2\pi i} \sum_{n=1}^{m+1} S_{n-1} \left\{ \sum_{i=1}^{n-1} \left[\frac{z^{i-1}}{n-1} (a^{n-1} - (-a)^{n-1}) \right] + z^{n-1} \ln \frac{z-a}{z+a} \right\} \tag{39}$$

By substituting this form of the function $\Phi(z)$ into relations (20) and (21), the corresponding equations of the caustics can be obtained.

However, besides these caustics formed at the ends of the load, another form of caustics is created by the deformed boundary, which is called a *pseudocaustic*. This caustic is expressed by:

$$W = \lambda_m z + 4C\overline{\Phi'(z)} \tag{40}$$

while the initial curve is expressed by:

$$z = x \tag{41}$$

By using relations (39) and (40–41), it is possible to determine the values of the unknown coefficients S_{n-1} in relation (39) and thus to determine the loading distribution along the line of contact. To achieve it, we have just to make a series of measurements on the experimentally obtained pseudocaustic and, afterwards to solve a system of linear equations.

Figure 17(b) presents the caustics and the pseudocaustics formed by the stress distribution of Fig. 17(a). By applying the above-mentioned

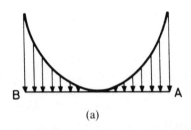

(a)

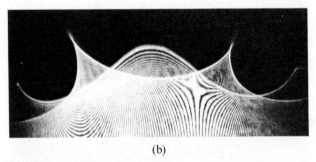

(b)

FIG. 17.   Load distribution (a) and the corresponding experimental caustics and pseudocaustics (b) by loading the semi-infinite plate.

numerical technique, the values of the load distribution of Fig. 17(a) at the six points indicated were evaluated by an approximation of the order of 5%.

The solution of this inverse problem where from the sizes and forms of the caustics and pseudocaustics the load distribution along the contact zone can be determined is interesting in the praxis and it cannot be solved with such an accuracy and sensitivity by any other experimental method. In the previous application the simplest case was encountered where the bodies in contact are in an elastic plane-stress condition and the boundary of either of them is a straight line.

Another case solved by the method of caustics is the case appearing regularly in the praxis of two cylindrical discs under elastic contact. Here the assumptions of Muskhelishvili were introduced of a load distribution at contact proportional to $x^{1/2}$ and it was assumed that the contact length is small compared to the radii of the cylinders. It was found that, under these conditions, the opening distance between the extremities of the caustic formed yields the unknown contact length, as well as the amount of the applied load for creating the contact.[44]

In the above-described cases only normal loads along the contact area were encountered. However, the method can be applied to the more realistic cases where two elastic bodies with arbitrary shapes of the boundaries under generalised plane-stress conditions are in contact and the loads transmitted from the one to the other through the contact are arbitrary, analysed in normal and tangential distributions. Moreover, due to the conditions of contact surfaces there are, assumed existing along the contact zones, jumps either of loading, or of their slope and curvature as well as jumps of displacement. All these jumps create singularities of various orders and these singularities may be studied by the forms and sizes of the caustics created there. Moreover, pseudocaustics yield all the necessary information for finding the load distribution at the contact zone outside singularities. All these problems are under development in our laboratory and some preliminary papers have already been presented.[45]

LATERALLY LOADED FLEXED PLATES[46–47]

While in the preceding problems plane specimens were subjected to in-plane two-dimensional stress systems, in the applications of this section loading systems normal to the plane of the specimens will be considered. For the study of flexed plates two possibilities exist. Either the caustic is formed from reflections of the impinging light beam all over the lateral faces of the plate, or the caustic is formed, as previously, by the reflected-light rays from a very small region restricted at the close vicinity of the singular stress field created at some discontinuity of the geometry of the plate during its loading. The first case concerns thin plates loaded within the elastic range, where deflections are kept small and there are no singularities in the stress field. For this case eqns. (8) may be applied for the determination of the caustic. In the second case the stress field of the flexed plate contains a singularity either due to the geometry of the plate (cracks or interfaces) or due to loading (concentrated loads). In this case the singular stress field is studied by following the same procedure as for the case of plane elastic problems subjected to in-plane forces, with the only difference that, instead of using Westergaard's complex function $\Phi(z)$ for evaluating k's the Sih expressions for the components of internal moments m_x, m_y and m_{xy} are used as they have been derived from the Reissner sixth-order plate-bending theory.[48] For the case of bending of plates the k-factors are called the k_1 and k_2 moment intensity factors, while k_3 is called shear force intensity factor. The appropriate equations expressing these intensity factors indicate that normal bending of the plate produces k_1, while parallel bending does not give rise to a moment singularity. Finally pure twisting influences both k_2 and k_3.[48] The procedure for evaluating moment and shear force intensity factors in bent plates is similar to that followed for plane-stress elastic problems. However, the form and variation of the initial curves and the caustics created are different from those for plane-stress problems.[49]

For the case of flexed plates which do not contain singularities it can readily be observed that relation (7) defines an initial curve on the plate for each position of the reference screen, placed at some distance z_0 from the plate. This curve is connected to the caustic on the reference screen through relations (8). Thus, by placing the reference screen at different distances z_0 from the plate, different caustics can be obtained from the reflected light rays from the corresponding initial curve on the plate. This correspondence between the initial curve on the plate and the caustic on the reference screen can be established experimentally, by tracing, for example, a fine grid on the plate and observing the images of points of the grid on the reference screen. Thus, for each point $P'(W_x, W_y)$ on the screen, the corresponding point $P(x, y)$ on the plate can be determined. If this correspondence between pairs of points on the specimen and the screen is established, relations (8) enable the direct calculation of the partial derivatives $\partial f(x, y)/\partial(x, y)$ at the point $P(x, y)$ on the plate. By placing the reference screen at various distances from the plate, the partial derivatives at all points of the plate can be determined. By a single differentiation of the obtained partial slopes of plate, its bending moments M_x, M_y and M_{xy} can be determined.[50]

Two particular cases were considered to show the importance of the above developments to the experimental solution of flexed plates and to provide an estimation of the accuracy of the method: (1) The simply supported equilateral triangular and square plates and (2) the general case of an axisymmetric plate. In the former case the theoretical and the experimental caustics of the plate were obtained and a comparison between them was made, while in the latter case the above-described technique leads to a direct experimental determination of the curvature of the plate.

For the case of the triangular plate loaded by a uniformly distributed load q, the deflection z of the plate is expressed by:

$$z = \frac{q}{64\alpha D}\left[x^3 - 3y^2x - \alpha(x^2 + y^2) + \frac{4}{27}\alpha^3\right]\left(\frac{4}{9}\alpha^2 - x^2 - y^2\right) \quad (42)$$

where α is the height of the triangle and the coordinates x and y are referred to a Cartesian system with its centre at the centre of gravity of the triangle and the x-axis coinciding with one height. By substituting relation (42) into eqn. (7) we obtain the following equation for the initial curve of the caustic:

$$\left[1 - 2C\left(r^3\cos 3\vartheta - \alpha r^2 + \frac{4\alpha^3}{27}\right)\right]^2$$

$$= C^2\left[\left(3r^2 - \frac{2\alpha^2}{3}\right)^2 r^2 + \alpha^2 r^4 - 2\alpha\left(3r^2 - \frac{2\alpha^2}{3}\right)r^3\cos 3\vartheta\right] \quad (43)$$

where:

$$C = \frac{z_0 q}{8\alpha D\lambda_m}$$

Equations of the caustic are defined by relations (8) by substituting the values of the partial derivatives of the deflection of the plate given by relations:

$$\frac{\partial z}{\partial x} = \frac{q}{64\alpha D}\left[-5x^4 + 4\alpha x^3 + 6\left(\frac{2\alpha^2}{9} + y^2\right)x^2\right.$$

$$\left. + 4\alpha\left(y^2 - \frac{8\alpha^2}{27}\right)x + 3y^2\left(y^2 - \frac{4\alpha^2}{9}\right)\right]$$

$$\frac{\partial z}{\partial y} = \frac{q}{64\alpha D}\left[4(3x + \alpha)y^3 + 4\left(x^3 + \alpha x^2 - \frac{2\alpha^2}{3}x - \frac{8\alpha^3}{27}\right)y\right] \quad (44)$$

From the thus defined equations of the caustics we can derive also the caustics formed by the boundary of the triangular plate as initial curve, which are again pseudocaustics, by putting, for example for the boundary with $x = -(\alpha/3)$, $y = 0$, these values into the equations of the caustics. Thus, we obtain the following equations:

$$W_x = \lambda_m\left[-\frac{\alpha}{3} + \frac{C}{4}\left(3y^4 - 2\alpha^2 y^2 + \frac{\alpha^4}{3}\right)\right]$$

$$W_y = \lambda_m y \quad (45)$$

with $-(\alpha/\sqrt{3}) \leq y \leq (\alpha/\sqrt{3})$.

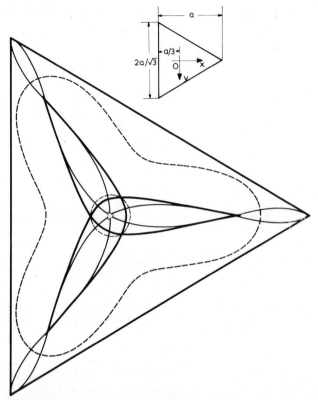

FIG. 18.   Theoretical forms of caustics and pseudocaustics obtained by loading a
triangular simply supported plate with a uniform load.

Figure 18 presents the caustics and pseudocaustics formed by loading a triangular simply supported plate with a uniform loading, while in Fig. 19 the corresponding experimental caustic is presented. It can be observed that the experimental caustic fits well with the theoretical one.

Similarly, the equations of the caustics formed by loading a simply supported square plate by a uniform load q can be obtained from relations (8).[47]

Finally, for the simple case of an axisymmetric plate, whose deflection curve is of the form $z = f(r)$, eqns. (7) and (8) yield:

$$W(r) = \lambda_m r - 2z_0 \frac{\mathrm{d}f(r)}{\mathrm{d}r} \qquad (46a)$$

FIG. 19.   Experimentally obtained caustics for the case of Fig. 18.

and:

$$\frac{d^2 f(r)}{dr^2} = \frac{\lambda_m}{2z_0} \tag{46b}$$

As it was already pointed out, it is a simple matter to determine the particular point (circle for the axisymmetric plate) on the plate, which corresponds to a given point of the caustic formed on the screen at a distance z_0 from the plate. This can be done, for example, by tracing a fine set of concentric circles on the plate and observing their reflections on the reference screen. After the determination of the corresponding points on the plate, whose plate curvature is given by the relation $\lambda_m/2z_0$, the distribution of the first derivative on the plate can be easily found by a numerical integration of the second derivatives, the constant of integration being calculated from the boundary conditions. After the determination of the first and the second derivatives of the deflection of the plate, the radial M_r and the tangential M_t components of moment on the plate can be easily determined by:

$$M_r = -D\left(\frac{d^2 f}{dr^2} + \frac{v}{r}\frac{df}{dr}\right) \qquad M_t = -D\left(v\frac{d^2 f}{dr^2} + \frac{1}{r}\frac{df}{dr}\right) \tag{47}$$

Besides the above-outlined procedure, the determination of the point on the plate at which the corresponding curvature is given by relation (46a) can be made by substituting the differential quantity dr in relation (45) by the differential quantity dz_0 of the distance z_0 between the plate and the reference screen Sc. Indeed, from relations (45) and (46b) it can be derived that:

$$r = \frac{1}{\lambda_m}\left(W - z_0\frac{dW}{dz_0}\right) \tag{48}$$

Relation (48) enables the experimental determination of point $P(r)$ on the plate, whose curvature is given by relation (46b). Indeed, the radius W of the caustic on the reference screen Sc at a distance z_0 from the plate can be directly determined experimentally, since the derivative dW/dz_0 can be approximated by:

$$\frac{dW}{dz_0} = \frac{\Delta W}{\Delta z_0} = \frac{W_1 - W_2}{\Delta z_0} \tag{49}$$

where W_1 and W_2 are the radii of the caustics obtained when the reference screen is moved by $\Delta z_0/2$ and $(-\Delta z_0/2)$ respectively from its position at a distance z_0 from the plate.

The exact location of point $P(r)$ on the plate allows the reference of the first derivative df/dr at this point and this derivative can be directly determined from relation (46) as follows:

$$\frac{df}{dr} = \frac{\lambda_m r - W}{2z_0} \tag{50}$$

CONCLUSIONS

In the present review paper the optical method of caustics was applied to the solution of the following six categories of engineering problems, which are of great importance in practical applications: (a) crack problems in elastic and elasto-plastic materials, (b) cracked shells, (c) dynamic crack propagation phenomena, (d) composite materials, (e) contact problems and (f) laterally loaded flexed plates. In each case the characteristic parameters of the corresponding stress fields developed were determined. These are the stress intensity factors and the order of the stress singularity in the cases of singular stress fields, as well as the determination of stress distributions, plastic zone lengths, bending moments, and generally the most characteristic quantities of the stress fields considered.

It was proved that this method is particularly convenient for the study of singular stress fields. In these cases the well-known methods of experimental stress analysis, such as photoelasticity, moiré, holography, strain gauges, etc. fail to yield reliable results. This is due to the steep thickness and refractive index variations of the specimen at the close vicinity of the singularity, which create an anomalous local region in the corresponding patterns of the aforesaid methods, difficult to analyse. Furthermore, holographic interferometry seems to be the most reliable method over all the conventional experimental stress-analysis methods. This method, which yields, among others, the thickness variation at the vicinity of the crack tip, contains fringes which are regularly spaced with a progressively increasing interfringe distance from the crack tip. In order to evaluate the stress intensity factor from these fringe patterns it is necessary to count the fringes at the vicinity of the crack tip, which is very difficult if not impossible. Thus, various extrapolation techniques for obtaining information from the far from the crack-tip region at the near region to the crack tip are of common use. However, extrapolation techniques cannot be accepted as accurate procedures in stress analysis.

On the other hand holographic patterns yielding very dense fringes around the crack tips fail to yield information at the region of singularity presenting black dots instead of fringes (see for instance the hologram of a cracked plexiglas plate given in Fig. 2 of reference 51) attributed to crazing phenomena when polymeric models are used. However, in a striking example given by reference 51 where a near field interferogram is given of the overall hologram, an internal caustic appears surrounding the dark dot at the crack tip.[52] This remark is of great importance when one compares the two methods. The method of caustics highly enlarges an infinitely small area at the close vicinity of the singular point and evaluates the order of singularity and the stress intensity factor from this area. The interferometric methods consider the same area as 'a factor limiting the extent of the meaningful data at the crack tip'. (Reference 51, p. 51.)

Therefore, the comparison of the method of caustics with the other experimental methods of stress analysis proves the potentialities and the superiority of the method of caustics for the study of singular stress fields.

REFERENCES

1. THEOCARIS, P. S. and GDOUTOS, E. E., *Appl. Optics*, 1976, **15**, 1629.
2. THEOCARIS, P. S. and GDOUTOS, E. E., *J. Phys. D. Appl. Phys.*, 1974, **7**, 472.

3. THEOCARIS, P. S., *J. Strain Anal.*, 1973, **8**, 267.
4. THEOCARIS, P. S., *Appl. Optics*, 1971, **10**, 2240.
5. MUSKHELISHVILI, N. I., *Some Basic Problems of the Mathematical Theory of Elasticity*, 2nd edn., Noordhoff, Groningen, 1963.
6. BORN, M. and WOLF, E., *Principles of Optics*, Pergamon Press, Oxford, 1970.
7. COKER, E. G. and FILON, L. N. G., *A Treatise on Photoelasticity*, Cambridge University Press, Cambridge, 1957.
8. THEOCARIS, P. S., *J. Appl. Mech.* (*Trans. ASME, Ser. E*), 1970, **37**, 409.
9. THEOCARIS, P. S., *Proc. Acad. Athens*, 1971, **46**, 116.
10. THEOCARIS, P. S., *Materialprüfung*, 1971, **13**, 264.
11. THEOCARIS, P. S., *Tekhnika Khronika*, 1971, **41**, 145.
12. THEOCARIS, P. S. and GDOUTOS, E. E., *J. Appl. Mech.* (*Trans. ASME, Ser. E*), 1972, **39**, 91.
13. THEOCARIS, P. S., *Exp. Mech.*, 1971, **11**, 280.
14. THEOCARIS, P. S. and IOAKIMIDES, N., *Zeit. ang. Math. Phys.*, 1971, **22**, 876.
15. THEOCARIS, P. S., *Proc. Int. Symp. Exp. Mech.*, Waterloo, Canada, June 1972, 511.
16. ASTM Special Technical Publication No. 381, *Fracture Toughness Testing and its Applications*, 1965.
17. THEOCARIS, P. S., *J. Strain Anal.*, 1972, **7**, 186.
18. THEOCARIS, P. S., *Int. J. Mech. Sci.*, 1972, **14**, 79.
19. THEOCARIS, P. S., *Acta Mech.*, 1973, **17**, 169.
20. THEOCARIS, P. S., *J. Mech. Phys. Solids*, 1972, **20**, 265.
21. THEOCARIS, P. S. and BLONZOU, C., *Materialprüfung*, 1973, **15**, 123.
22. THEOCARIS, P. S., *Proc. 3rd Int. Conf. Fracture*, München, 1973, Vol. VI, paper No. 512.
23. THEOCARIS, P. S., *Int. J. Mech. Sci.*, 1974, **16**, 855.
24. THEOCARIS, P. S., *J. Strain Anal.*, 1974, **9**, 197.
25. THEOCARIS, P. S., *Appl. Optics*, 1973, **12**, 380.
26. THEOCARIS, P. S., *Int. J. Fract.*, 1973, **9**, 185.
27. THEOCARIS, P. S. and GDOUTOS, E. E., *Engrg. Fract. Mech.*, 1974, **6**, 523.
28. THEOCARIS, P. S. and GDOUTOS, E. E., *Int. J. Fract.*, 1974, **10**, 549.
29. THEOCARIS, P. S., *Int. J. Mech. Sci.*, 1975, **17**, 475.
30. THEOCARIS, P. S., *Experimental Mechanics*, 1963, **3**, 207.
31. ROSENFIELD, A. R., DAI, P. K. and HAHN, G. I., *Proc. 1st Int. Conf. Fracture*, Sendai, Japan, 1966, **1**, 223.
32. THEOCARIS, P. S., *Proc. First Int. Conf. Struct. Mech. Reactor Techn.*, Berlin, 1972, Vol. 4, pp. 487–505.
33. THEOCARIS, P. S. and THIREOS, C., *Int. J. Fract.*, 1976, **12**, 691.
34. KATSAMANIS, F., RAFTOPOULOS, D. and THEOCARIS, P. S., *J. Engrg. Mat. Techn.*, 1977, **99**, 105.
35. RAFTOPOULOS, D., KATSAMANIS, F. and THEOCARIS, P. S., *Exp. Mech.*, 1977, **17**, 128.
36. THEOCARIS, P. S. and KATSAMANIS, F., *Engrg. Fract. Mech.*, 1978, **10**, 256.
37. IRWIN, G. R., Fracture, in *Encyclopedia of Physics* (Flügge, S., ed.), Springer-Verlag, Berlin, 1958.
38. THEOCARIS, P. S., *Zeit. ang. Math. Phys.*, 1975, **26**, 77.
39. THEOCARIS, P. S., *Acta Mech.*, 1976, **24**, 99.

40. THEOCARIS, P. S. and STASSINAKIS, C. A., *Int. J. Fract.*, 1977, **13**, 13.
41. THEOCARIS, P. S., *Int. J. Solids Struct.*, 1973, **9**, 655.
42. THEOCARIS, P. S. and RAZEM, C., *J. Appl. Mech.* (*Trans. ASME, Ser. E*), 1978, **45** (in press).
43. THEOCARIS, P. S. and RAZEM, C., *J. Strain Anal.*, 1977, **12**, 223.
44. THEOCARIS, P. S. and STASSINAKIS, C. A., *Exp. Mechanics*, 1978, **18**, 329.
45. THEOCARIS, P. S., *Proc. Third Bulg. Nat. Cong. Theor. Appl. Mech.*, Varna, Bulgaria, 1977, **1**, 263.
46. THEOCARIS, P. S. and GDOUTOS, E. E., *J. Appl. Mech.* (*Trans. ASME, Ser. E*), 1977, **44**, 107.
47. THEOCARIS, P. S., *Int. J. Solids Struct.*, 1977, **13**, 1281.
48. SIH, G. C., *Plates and Shells with Cracks* (*Mechanics of Fracture*, 3), 1st edn., Noordhoff, Leyden, 1977.
49. THEOCARIS, P. S., *Exp. Mechanics* (submitted for publication).
50. TIMOSHENKO, S. and WOINOWSKY-KRIEGER, S., *Theory of Plates and Shells*, 2nd edn., McGraw-Hill, New York, 1959.
51. DUDDERAR, T. D. and O'REGAN, *Exp. Mech.*, 1971, **11**, 49.
52. THEOCARIS, P. S., *Exp. Mech.*, 1975, **15**, 150.

Chapter 3

MODERN METHODS IN POLARISATION OPTICS

P. S. THEOCARIS

The National Technical University, Athens, Greece

SUMMARY

The necessity of solving polarisation optics problems involving the passage of an elliptically polarised light beam through a pile of optical elements in a rapid, unique, elegant and efficient way led to the introduction of the modern methods of polarisation optics based on the Poincaré sphere method and the Mueller and Jones calculi. These methods for describing the various forms of elliptical polarisation may be classified into two groups, that is: analytic methods and graphical methods. The analytic methods including also all numerical evaluations of the elements contained in the calculations are based on either the Mueller, or the Jones calculi. To the two main analytic methods based on Mueller and Jones calculi another one, the method of quaternions, was developed which is based on Jones calculus where any Jones matrix is expressed by a linear combination of the form Pauli fundamental matrices.

The graphical methods are based on the concept of Poincaré sphere. According to this method a polarisation form is defined by a corresponding point on the surface of the sphere. All transformations of polarisation forms due to the passage of light through an optical element takes place on the surface of the Poincaré sphere. Two projections of the sphere one parallel to its north–south axis on its equatorial plane creates the j-circle method, while a stereographic projection of the sphere on its principal meridian yields the so-called Wulff-net method.

In this chapter we use these modern methods to describe the influence of a pile of optical elements on the state of polarisation of an incident elliptically polarised monochromatic beam. This most general problem is on the basis of plane and three-dimensional photoelasticity. While this paper will deal with cases of problems of polarisation optics it will not enter into problems of photoelasticity and to its natural extension, that is holography and holographic interferometry. This part of study will make the subject of a companion paper.

Generally, in this chapter we limit ourselves to developing general principles for the solution of problems of polarisation optics without entering into the details of such solutions.

INTRODUCTION

Photoelasticity from its introduction in the early 1930s up to the present has seen a tremendous development in its use for the solution of a large variety of problems in experimental stress analysis. The method is based on the Neumann–Maxwell stress–optical law. All applications of photoelasticity in stress analysis were based on classical techniques of vector analysis and analytic geometry using the conventional light vector concept. This classical way of solving problems of mechanics has indicated many shortcomings and drawbacks especially when three-dimensional elasticity problems had to be treated and in several other cases with problems pertaining to complicated load distributions and geometric situations.

Concurrently with the above-mentioned conventional methods of treating problems of mechanics by photoelasticity a series of efficient techniques have come into prominence, transplanted from similar methods used in crystallography, electric-circuit theory, geodesy, etc.

These new techniques are concerned with predicting the effects of polarisers and retardation plates and may be classified into two categories, i.e. analytic techniques and graphical methods. Both these types of techniques reduce what formerly seemed a hopelessly difficult problem of evaluating the behaviour of any combination of a rotator and a retarder to a simple problem. The analytic methods developed for treating problems of the behaviour of retarders and rotators are based on the Stokes and Jones vectors, which are already used extensively in crystallography. Attached to the Stokes vector which was introduced in 1852[1] a calculus was developed by Mueller in 1948,[2] the so-called Mueller calculus, for treating phenomena of polarised light propagation. The Jones vector and the corresponding Jones calculus were conceived by Jones in 1941.[3-9] The effectiveness of these analytic methods is based on the simplicity and the consistency in predicting the effect of an optical element, such as a polariser or a retarder, on a beam of polarised light passing through this element.

In the Mueller calculus a light beam is represented by a four-element column vector, the Stokes vector, and each optical element by a four-by-four real matrix, while the Stokes vector for the light emerging from an optical device is given by the matrix multiplication of the incident Stokes vector by the matrix of the optical element. A similar procedure is valid for the Jones calculus with the difference that, a polarised-light beam is represented by a two-element complex column matrix and each optical device by a two-by-two complex matrix.

To these two basic calculi for evaluating the behaviour of an optical element a third method should be added based on the quaternions and using the so-called Pauli spin matrices.[10] According to this method the unitary Jones matrices for retarders are expressed in terms of the four basic Pauli spin matrices. The concept of quaternion was first introduced to photoelasticity by Richartz and Hsu[11] who realised that the coefficients of the quaternion completely determine the corresponding optical element. The method of quaternions was widely used by Cernosek[12-14] for the solution of problems of polarisation optics. He also presented a graphical method for the study of such problems which was based on the theory of quaternions.[15]

Alongside the analytic methods, graphical methods have also been developed. These methods are based on the properties of the Poincaré sphere[16] used for representing polarisation profiles of light beams emerging from any optical element (retarder or rotator). According to Poincaré's sphere representation a polarisation state is represented by a point on a unit-radius sphere and the result of an optical element on the polarised light is given by a suitable transformation on the surface of the sphere. Thus, the problem of finding the emerging form of polarised light from an optical element can be solved graphically by a suitable displacement on the Poincaré sphere.

The Poincaré sphere method was applied by many investigators to problems of polarisation optics at the beginning of this century.[17-22] Jerrard[23] gave a brief account of the theory of Poincaré sphere, while Koester[24] gave suitable spheres with convenient degree scales. Ramachandran and Ramaseshan[25] presented a comprehensive and thorough discussion of polarisation optics problems connected with the Poincaré sphere. Pancharatnam[26-27] used the Poincaré sphere for the study of the interference of two coherent light beams and the resolution of any polarisation form into two given components. For a thorough listing of references related to applications of the Poincaré sphere in polarisation optics and especially in photoelasticity the reader is referred to references 28 and 29.

Instead of working on the Poincaré sphere, which is a three-dimensional device, it is possible to project the sphere into a plane. Two such projections have been already developed. The first one corresponds to a parallel projection of the sphere to its equatorial plane and solves the problem by working with the projection vector of each point of the sphere on the equatorial plane; the so-called j-vector.[30-32] In this case each polarisation form is represented by an interior point in the projection, which is called the j-circle. The influence of an optical element on an impinging polarisation form consists in angularly displacing and changing the size of the initial j-vector.

The other type of projection of the Poincaré sphere exists where a projecting pole is chosen as one of the intersections of the normal to the principal meridional and the equator, and a projecting plane as the plane of the principal meridional of the sphere. In this way the one hemisphere containing the pole is projected on the exterior of the meridian, while the other hemisphere is projected on the interior. In this stereographic type of projection the meridional and parallel circles are projected as circles, and the equator and the normal to the principal meridional as diameters perpendicular to each other. To this type of projection a Wulff-type net is drawn, which facilitates the finding of angular coordinates of the projection. This Wulff-net stereographic projection maintains the angles and facilitates the calculations. The idea of using such a projection came from geodesy and navigation charts and was first applied to crystallography and petrography. An early application of stereographic projection for solving polarisation problems was made by Wright[33] who gave a complete spectrum of applications of the Poincaré sphere as well as an authoritative study of the Wulff-net stereographic projection to polarisation optics problems. Other researchers who applied the Poincaré sphere and the Wulff-net stereographic projection methods to classical problems of photoelasticity were Robert,[34] Aben[35-36] and Schwieger.[37]

All these methods of describing a polarisation form are characterised by a particular elegance and they provide a powerful means for dealing with problems of polarisation optics especially in three-dimensional photoelasticity where a short-cut solution may be obtained by dividing the model into a pile of successive optical elements and applying to each of them the corresponding calculus.

It is the purpose of this chapter to outline and review these novel and powerful methods, presenting them in a unified mode for the description of any form of polarised light passing through a rotator and/or retarder. In this way complicated problems of two- and three-dimensional photoelasticity may be solved in a straightforward and comprehensive manner and simple and elegant solutions may be devised for complex problems which, because of their particular nature, were not solved by classical methods.

VECTORIAL REPRESENTATION OF POLARISED LIGHT

Vector calculus may be used to describe the light vector, which represents the electric-field vector according to the electromagnetic theory of light, or the position vector of the ether particle at each instant according to the wave theory of light. We now express vectorially the linear circular and elliptical polarisation forms of light.

In the more general case of an elliptical polarisation the light vector $\mathbf{a}$ is expressed by:

$$\mathbf{a} = \alpha_x \mathbf{i} + \alpha_y \mathbf{j} \tag{1}$$

with:

$$a_x = A_x \text{Re}\{\exp[(v + \delta_x)i]\} \quad \text{and} \quad a_y = A_y \text{Re}\{\exp[(v + \delta_y)i]\} \tag{2}$$

In these relations A_x, A_y are the components of the amplitude A of the light vector $\mathbf{a}$ along the axes of the reference frame $Oxyz$ for which the z-direction is the direction of the propagation of light and Oxy a transverse plane passing through the origin O. Moreover, v expresses the quantity given by $v = (\omega t + 2\pi z/\lambda)$, λ is the wavelength of the monochromatic light used, ω is the angular frequency given by $\omega = 2\pi/T$ where T is the period of the oscillation, and $\mathbf{i}$, $\mathbf{j}$ unit-vectors along the Ox- and Oy-axes respectively. Finally, δ_x and δ_y are the components of phase along the Ox- and Oy-axes with a phase difference $\delta = (\delta_y - \delta_x)$.

Equations (1) and (2) represent the parametric equations of an ellipse. Indeed, they can be further transformed into the relations:

$$\frac{\alpha_x}{A_x} \sin \delta_y - \frac{\alpha_y}{A_y} \sin \delta_x = \cos v \sin \delta$$

$$\frac{\alpha_x}{A_x} \cos \delta_y - \frac{\alpha_y}{A_y} \cos \delta_x = \sin v \sin \delta \tag{3}$$

which further yield, by squaring and adding them, the equation:

$$\left(\frac{\alpha_x}{A_x}\right)^2 + \left(\frac{\alpha_y}{A_y}\right)^2 - \frac{2\alpha_x \alpha_y}{A_x A_y} \cos \delta - \sin^2 \delta = 0 \tag{4}$$

which expresses in general a second degree curve in the α_x, α_y − plane. But since the quantity:

$$\left(\frac{1}{A_x}\right)^2 \left(\frac{1}{A_y}\right)^2 - \frac{\cos^2 \delta}{A_x^2 A_y^2} = \left(\frac{\sin \delta}{A_x A_y}\right)^2 \tag{5}$$

is always positive, eqn. (4), whose parametric form is given by relations (3), always represents an ellipse indicated in Fig. 1.

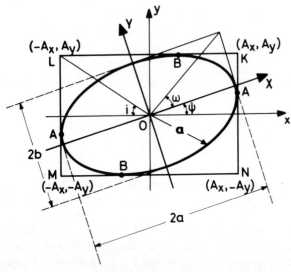

FIG. 1.   Notation of the various quantities entered in an elliptically polarised light.

The maxima and minima for the components α_x, α_y of the vector **a** describing the ellipse can be readily found from eqn. (4) and they are points A and B in Fig. 1 with coordinates:

$$\alpha_x^A = \pm A_x \qquad\qquad \alpha_x^B = \pm A_x \cos \delta$$
$$\alpha_y^A = \pm A_y \cos \delta \qquad \alpha_y^B = \pm A_y \tag{6}$$

From eqns. (6) it can be derived that the ellipse expressed by eqn. (4) is inscribed into the rectangle KLMN with sides parallel to the coordinate axes and equal to $\text{KL} = \text{MN} = 2A_x$ and $\text{KN} = \text{LM} = 2A_y$.

The principal axes of the ellipse with respective lengths $2a$ and $2b$ ($a \geq b$) are along the new reference frame OXY, subtending an angle ψ with the

Oxy-frame, which is called the azimuth. It can be readily found that the following relations are valid:[38]

$$a^2 + b^2 = A_x^2 + A_y^2 \tag{7}$$

$$\pm \, \alpha b = A_x A_y \sin \delta \tag{8}$$

and:

$$\tan 2\psi = \frac{2A_x A_y}{A_x^2 - A_y^2} \cos \delta \tag{9}$$

Introducing the angle i such that:

$$\tan i = \frac{A_y}{A_x} \tag{10}$$

eqn. (9) takes the form:

$$\tan 2\psi = \tan 2i \cos \delta \tag{11}$$

Equation (11) defines the azimuth ψ of the light ellipse.

On the other hand, relations (7) and (8) divided by parts yield:

$$\pm \frac{2ab}{a^2 + b^2} = \frac{2A_x A_y}{A_x^2 + A_y^2} \sin \delta$$

which, by introducing the angle ω expressed by:

$$\tan \omega = \pm \frac{b}{a} \tag{12}$$

becomes:

$$\sin 2\omega = \sin 2i \sin \delta \tag{13}$$

Angle ω, defined by relation (12), is called the *ellipticity* of the light ellipse.

On the other hand the difference $(a^2 - b^2)$ is expressed by:[38]

$$(a^2 - b^2) = (A_x^2 - A_y^2)(\cos^2 \psi - \sin^2 \psi) + 2A_x A_y \sin 2\psi \cos \delta$$

which by taking into account relations (7) and (11) yields:

$$\cos 2i = \cos 2\omega \cos 2\psi \tag{14}$$

Equations (11), (13) and (14) constitute the fundamental relations connecting the four angular quantities ψ, ω, i and δ.

Relations (7), (8), (11), (13) and (14) relate the principal semi-axes a and b of the light ellipse and its azimuth with the amplitudes A_x, A_y and the phase difference δ of the original vibrations of polarised light.

The appearance in the above relations of the plus and minus signs indicates the two opposite directions in which the light vector is describing the light ellipse (left-handed and right-handed polarisation).

According to this definition an elliptical polarisation is called right-handed when the observer looking towards the light source sees the light vector moving in the clockwise sense. It can be readily shown from the above equations that for a right-handed elliptical polarisation it is valid that $\sin \delta > 0$, that is, $0° < \delta < 180°$, while for a left-handed elliptical polarisation, $180° < \delta < 360°$.

For the case when the phase difference δ takes the values $\delta = 0°$, $180°$ and $360°$, the light ellipse degenerates into a straight line and we have the case of a linear polarisation, while in the case when $A_x = A_y$ and $\delta = 90°$ the elliptical polarisation becomes a circular polarisation. In this case the ellipse inscribed into the rectangle KLMN becomes a circle and the rectangle degenerates into a square ($2a = 2b, 2A_x = 2A_y$).

From the above brief study it can be readily derived that the light ellipse in an elliptically polarised light is characterised by three independent quantities that is; either the amplitudes A_x and A_y of the original vibrations and their phase difference δ, or by the lengths of the principal semi-axes a and b of the light ellipse and its azimuth ψ. However, in order to completely define the light ellipse, it is necessary to indicate also the direction of rotation of the light vector describing the ellipse. Since, as was previously indicated, for a right-handed polarisation it is valid that $0° < \delta < 180°$ and for a left-handed polarisation it is valid that $180° < \delta < 360°$, it can be derived from relation (13) that the ellipticity ω should vary between $0° < \omega < 45°$ for a right-handed polarisation and between $-45° < \omega < 0°$ for a left-handed polarisation.

In many problems including the study of the passage of polarised light through a train of optical elements we are not interested in the absolute value of the light intensity $I = (a^2 + b^2) = (A_x^2 + A_y^2)$ but in its relative variation, that is; we are interested in the change of shape of the light ellipse. In this way we can normalise the light vector to a unit intensity ($I = 1$) and thus the angles ω ($-45° \leq \omega \leq 45°$) and ψ ($0° \leq \omega < \pi$) can completely define the normalised light ellipse.

ANALYTIC REPRESENTATION OF POLARISATION FORMS

The Stokes Vector

The Stokes vector is a four-element column vector **S** which completely defines an elliptical polarisation form. The four elements of the vector are expressed by:

$$\mathbf{S} = \begin{bmatrix} s_0 \\ s_1 \\ s_2 \\ s_3 \end{bmatrix} = \begin{bmatrix} A_x^2 + A_y^2 \\ A_x^2 - A_y^2 \\ 2A_x A_y \cos \delta \\ 2A_x A_y \sin \delta \end{bmatrix} \tag{15}$$

It is clear that s_0 represents the intensity of polarised light, while all the other elements have dimensions of intensity. The form parameters s_0, s_1, s_2, s_3 are not independent, but they satisfy the identity:

$$s_0^2 = s_1^2 + s_2^2 + s_3^2 \tag{16}$$

Relations (7), (8) and (10) to (13) introduced into the Stokes vector elements yield:

$$s_1 = s_0 \cos 2\omega \cos 2\psi$$

$$s_2 = s_0 \cos 2\omega \sin 2\psi$$

$$s_3 = s_0 \sin 2\omega \tag{17}$$

The elements of the Stokes vector corresponding to a given polarisation form can be readily determined by using relations (15) and (17). Thus, for a linear and horizontally polarised light ($A_x \neq 0, A_y = \delta = 0$, or $\psi = \omega = 0$) we have:

$$s_0 = s_1 = A_x^2 \quad \text{and} \quad s_2 = s_3 = 0 \tag{18a}$$

For a right-circularly polarised light we have $A_x = A_y = A$ and $\delta = 90°$ or $\omega = 45°$ and therefore:

$$s_0 = s_3 = 2A^2 \quad \text{and} \quad s_1 = s_2 = 0 \tag{18b}$$

For a left-circularly polarised light we have $A_x = A_y = A$ and $\delta = -90°$ or $\omega = -45°$. Thus we obtain:

$$s_0 = 2A^2 \quad s_3 = -2A^2 \quad \text{and} \quad s_1 = s_2 = 0 \tag{19}$$

Usually we are interested only in the relative values of Stokes parameters normalised to the first parameter s_0. In this case these parameters correspond to a polarised light of unit intensity. The normalised Stokes vectors to s_0 for these particular cases are expressed by:

$$\mathbf{S}_{LH} = \begin{vmatrix} 1 \\ 1 \\ 0 \\ 0 \end{vmatrix} \quad \mathbf{S}_{CR} = \begin{vmatrix} 1 \\ 0 \\ 0 \\ 1 \end{vmatrix} \quad \text{and} \quad \mathbf{S}_{CL} = \begin{vmatrix} 1 \\ 0 \\ 0 \\ -1 \end{vmatrix}$$

Since all elements of the Stokes vector have dimensions of intensity, which is a scalar quantity, it can be readily deduced that the addition of two incoherent-light beams with Stokes vectors $\mathbf{S}'$ and $\mathbf{S}''$ respectively yield another light beam with Stokes vector $\mathbf{S}$ given by:

$$\mathbf{S} = \mathbf{S}' + \mathbf{S}'' = \begin{vmatrix} s'_0 \\ s'_1 \\ s'_2 \\ s'_3 \end{vmatrix} + \begin{vmatrix} s''_0 \\ s''_1 \\ s''_2 \\ s''_3 \end{vmatrix} = \begin{vmatrix} s'_0 + s''_0 \\ s'_1 + s''_1 \\ s'_2 + s''_2 \\ s'_3 + s''_3 \end{vmatrix} \tag{20}$$

This additive property of Stokes vectors facilitates the calculations related to the addition of two incoherent-light beams.

Thus, in conclusion, a unique correspondence between a given polarisation form and the respective Stokes vector is established through relations (15) and (17). Table 1 summarises all expressions of Stokes vectors corresponding to various types of polarised-light forms.

The Jones Vector

The Jones vector is a two-element complex column vector whose elements are equal to the a_x, and a_y components of the light vector $\mathbf{a}$. Thus, the Jones vector for an elliptical polarisation form is expressed by:

$$\mathbf{a} = \begin{vmatrix} a_x \\ a_y \end{vmatrix} = \begin{vmatrix} A_x \exp[i(v + \delta_x)] \\ A_y \exp[i(v + \delta_y)] \end{vmatrix}$$

Extracting the time factor $\exp(iv)$ from the above matrix we have

$$\mathbf{a} = \exp(iv) \begin{vmatrix} A_x \exp(i\delta_x) \\ A_y \exp(i\delta_y) \end{vmatrix}$$

Normalising now the Jones vector $\mathbf{a}$ to the time factor $\exp(iv)$ we obtain the simpler form of the Jones vector:

$$\mathbf{a} = \begin{bmatrix} A_x \exp(i\delta_x) \\ A_y \exp(i\delta_y) \end{bmatrix} \tag{21}$$

Usually the Jones vector may be normalised to whatever quantity is suitable to reduce the value of the corresponding light intensity to unity. Thus, another normalised form of expression (21) for the Jones vector is given by:

$$\mathbf{a} = \begin{bmatrix} \cos B \exp(-i\delta/2) \\ \sin B \exp(i\delta/2) \end{bmatrix} \tag{22}$$

TABLE 1
STOKES VECTORS IN POLARISED LIGHT

Elliptically polarised light	Circularly polarised light		Linearly polarised light		
	Right	Left	General at angle θ	Horizontal $\theta=0$	Vertical $\theta=(\pi/2)$
Light vector $\mathbf{a}$ $\mathbf{a} = a_x\mathbf{i} + a_y\mathbf{j}$ $a_x = A_x e^{i(v+\delta_x)}$ $a_y = A_y e^{i(v+\delta_y)}$ $\delta = \delta_y - \delta_x$ Light ellipse with azimuth ψ and ellipticity ω $(\tan\omega = b/a)$ b, a semi-axes of the ellipse.	$A_x = A_y = A$, $\delta = (\pi/2)$ $\omega = (\pi/4)$	$A_x = A_y = A$, $\delta = -(\pi/2)$ $\omega = -(\pi/4)$	θ with the Ox-axis $\delta = 0$, $(A_x/A) = \cos\theta$, $(A_y/A) = \sin\theta$, $\omega = 0$		

STOKES VECTORS

Elliptically polarised light	Circularly polarised light		Linearly polarised light		
$\begin{bmatrix} A_x^2 + A_y^2 \\ A_x^2 - A_y^2 \\ 2A_xA_y\cos\delta \\ 2A_xA_y\sin\delta \end{bmatrix}$ $\begin{bmatrix} s_0 \\ s_0\cos2\omega\cos2\psi \\ s_0\cos2\omega\sin2\psi \\ s_0\sin2\omega \end{bmatrix}^{(*)}$	$\begin{bmatrix} 2A^2 \\ 0 \\ 0 \\ 2A^2 \end{bmatrix}$	$\begin{bmatrix} 2A^2 \\ 0 \\ 0 \\ -2A^2 \end{bmatrix}$	$\begin{bmatrix} A^2 \\ A^2\cos2\theta \\ A^2\sin2\theta \\ 0 \end{bmatrix}$	$\begin{bmatrix} A^2 \\ A^2 \\ 0 \\ 0 \end{bmatrix}$	$\begin{bmatrix} A^2 \\ -A^2 \\ 0 \\ 0 \end{bmatrix}$

NORMALISED STOKES VECTORS

Elliptically polarised light	Circularly polarised light		Linearly polarised light		
$\begin{bmatrix} 1 \\ \cos2\omega\cos2\psi \\ \cos2\omega\sin2\psi \\ \sin2\omega \end{bmatrix}$	$\begin{bmatrix} 1 \\ 0 \\ 0 \\ 1 \end{bmatrix}$	$\begin{bmatrix} 1 \\ 0 \\ 0 \\ -1 \end{bmatrix}$	$\begin{bmatrix} 1 \\ \cos2\theta \\ \sin2\theta \\ 0 \end{bmatrix}$	$\begin{bmatrix} 1 \\ 1 \\ 0 \\ 0 \end{bmatrix}$	$\begin{bmatrix} 1 \\ -1 \\ 0 \\ 0 \end{bmatrix}$

$(*) s_0 = A_x^2 + A_y^2 = a^2 + b^2$

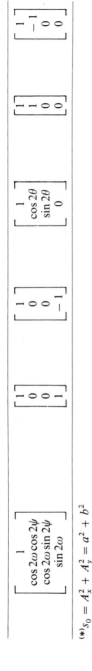

where $B = |\arctan A_y/A_x|$ and $\delta = (\delta_y - \delta_x)$. Relations (21) and (22) may be used indifferently to define any polarised light form. Thus, for a linear horizontally polarised light, for a right-circularly polarised light form and for a left-circularly polarised light form Jones vectors are expressed by one of these relations, if we apply relation (21):

$$\mathbf{a} = \begin{bmatrix} A_x \exp(i\delta_x) \\ 0 \end{bmatrix}, \mathbf{a} = \begin{bmatrix} A \exp(i\delta_x) \\ A \exp[i(\delta_x + \pi/2)] \end{bmatrix}$$

$$\text{and} \qquad \mathbf{a} = \begin{bmatrix} A \exp[i(\delta_x + \pi/2)] \\ A \exp(i\delta_x) \end{bmatrix}$$

and by taking $A_x = 1$, $A = (1/\sqrt{2})$ $\delta_x = 0$, $\mp \pi/2$ and $\delta_y = 0$ we obtain the normalised vectors for the above expressions such as to reduce the intensity to unity and also to reduce the vectors to simplest form as given by:

$$\mathbf{a} = \begin{bmatrix} 1 \\ 0 \end{bmatrix}, \mathbf{a} = 1/\sqrt{2} \begin{bmatrix} -i \\ 1 \end{bmatrix} \quad \text{and} \quad \mathbf{a} = 1/\sqrt{2} \begin{bmatrix} i \\ 1 \end{bmatrix}$$

Now, if we apply relation (22) the corresponding Jones vectors for the same particular cases examined above are expressed by:

$$\mathbf{a} = \begin{bmatrix} 1 \\ 0 \end{bmatrix}, \mathbf{a} = \frac{1+i}{2} \begin{bmatrix} -i \\ 1 \end{bmatrix} \quad \text{and} \quad \mathbf{a} = \frac{1-i}{2} \begin{bmatrix} i \\ 1 \end{bmatrix}$$

From either of relations (21) and (22) it can be seen that for a given polarisation form there is always a Jones vector corresponding to it and vice versa.

The complex conjugate $\tilde{\mathbf{a}}$ of the transpose of the vector $\mathbf{a}$, which is called the Hermitian conjugate matrix of $\mathbf{a}$ is given by:

$$\tilde{\mathbf{a}} = [A_x \exp(-i\delta_x) A_y \exp(-i\delta_y)]$$

then the light intensity I is given by:

$$I = \tilde{\mathbf{a}}\mathbf{a} = [A_x \exp(-i\delta_x) A_y \exp(-i\delta_y)] \begin{bmatrix} A_x \exp(i\delta_x) \\ A_y \exp(i\delta_y) \end{bmatrix} = A_x^2 + A_y^2 \quad (23)$$

As the Stokes vector is especially convenient for the addition of two incoherent-light beams, so the Jones vector is suitable for the addition of two coherent polarised-light beams. To achieve this one has to write down the Jones vectors of the two combined beams and add them. Indeed, if the two coherent-light beams are expressed by Jones vectors $\mathbf{a}'$ and $\mathbf{a}''$ the Jones

TABLE 2

JONES VECTORS IN POLARISED LIGHT

Elliptically polarised light	Circularly polarised light		Linearly polarised light		
	Right $A_x = A_y = A$ $\delta = (\pi/2)$	**Left** $A_x = A_y = A$ $\delta = -(\pi/2)$	General at an angle θ with the Ox-axis, $\delta = 0$	Horizontal $\theta = 0$ $A_y = 0$	Vertical $\theta = (\pi/2)$ $A_x = 0$
Light vector $\mathbf{a}$ $\mathbf{a} = a_x\mathbf{i} + a_y\mathbf{j}$ $a_x = A_x e^{i\delta_x}$ $a_y = A_y e^{i\delta_y}$ $\delta = \delta_y - \delta_x$					

JONES VECTORS

$\begin{bmatrix} A_x e^{i\delta_x} \\ A_y e^{i\delta_y} \end{bmatrix}$	$\begin{bmatrix} Ae^{i\delta_x} \\ Ae^{i(\delta_x + \pi/2)} \end{bmatrix}$	$\begin{bmatrix} Ae^{i(\delta_x + \pi/2)} \\ Ae^{i\delta_x} \end{bmatrix}$	$\begin{bmatrix} A_x e^{i\delta_x} \\ \pm A_y e^{i\delta_x} \end{bmatrix}$	$\begin{bmatrix} A_x e^{i\delta_x} \\ 0 \end{bmatrix}$	$\begin{bmatrix} 0 \\ A_y e^{i\delta_x} \end{bmatrix}$

NORMALISED JONES VECTORS

$\begin{bmatrix} \cos B\, e^{-i(\delta/2)} \\ \sin B\, e^{i(\delta/2)} \end{bmatrix}$	$\dfrac{1+i}{2}\begin{bmatrix} -i \\ 1 \end{bmatrix} \dfrac{1}{\sqrt{2}}\begin{bmatrix} -i \\ 1 \end{bmatrix}$	$\dfrac{1-i}{2}\begin{bmatrix} i \\ 1 \end{bmatrix} \dfrac{1}{\sqrt{2}}\begin{bmatrix} i \\ 1 \end{bmatrix}$	$\begin{bmatrix} \cos\theta \\ \pm\sin\theta \end{bmatrix}$	$\begin{bmatrix} 1 \\ 0 \end{bmatrix}$	$\begin{bmatrix} 0 \\ 1 \end{bmatrix}$

$$B = \left| \arctan\frac{A_y}{A_x} \right|$$

vector **a** of their combination is given by:

$$\mathbf{a} = \mathbf{a}' + \mathbf{a}'' = \begin{bmatrix} a'_x \\ a'_y \end{bmatrix} + \begin{bmatrix} a''_x \\ a''_y \end{bmatrix} = \begin{bmatrix} a'_x + a''_x \\ a'_x + a''_y \end{bmatrix}$$

Table 2 gives the Jones vectors corresponding to the most common types of polarisation forms.

THE METHOD OF QUATERNIONS

The analytic method of quaternions, which is based on Jones calculus, is also a powerful technique for the solution of photoelasticity problems. It can be derived that the general two-by-two complex Jones matrix can be expressed as a linear combination of the four Pauli fundamental matrices. Consequently, the coefficients of the quaternion of these matrices may completely determine the corresponding optical element.

Let us consider the general unitary matrix expressed by:

$$\mathbf{J} = \begin{bmatrix} A_0 + iA_1 & A_2 + iA_3 \\ -A_2 + iA_3 & -A_0 - iA_1 \end{bmatrix} \tag{24}$$

where the real numbers $A_i (i = 1, 2, 3, 4)$ satisfy the relation:

$$A_0^2 + A_1^2 + A_2^2 + A_3^2 = 1 \tag{25}$$

and the four Pauli matrices $\mathbf{Q}_0$, $\mathbf{Q}_1$, $\mathbf{Q}_2$, and $\mathbf{Q}_3$ which are given by:

$$\mathbf{Q}_0 = \begin{bmatrix} 1 & 0 \\ 0 & 1 \end{bmatrix}, \mathbf{Q}_1 = \begin{bmatrix} i & 0 \\ 0 & -i \end{bmatrix} \mathbf{Q}_2 = \begin{bmatrix} 0 & 1 \\ -1 & 0 \end{bmatrix} \quad \text{and} \quad \mathbf{Q}_3 = \begin{bmatrix} 0 & i \\ i & 0 \end{bmatrix} \tag{26}$$

Matrix **J** may be expressed as a quaternion as follows:

$$\mathbf{J} = A_0\mathbf{Q}_0 + A_1\mathbf{Q}_1 + A_2\mathbf{Q}_2 + A_3\mathbf{Q}_3 \tag{27}$$

It is further valid between matrices $\mathbf{Q}_i$ that:

$$\mathbf{Q}_1^2 = \mathbf{Q}_2^2 = \mathbf{Q}_3^2 = -\mathbf{Q}_0 \quad \text{and} \quad \mathbf{Q}_2\mathbf{Q}_3 = -\mathbf{Q}_3\mathbf{Q}_2 = \mathbf{Q}_1$$

$$\mathbf{Q}_3\mathbf{Q}_1 = -\mathbf{Q}_1\mathbf{Q}_3 = \mathbf{Q}_2 \quad \text{and} \quad \mathbf{Q}_1\mathbf{Q}_2 = -\mathbf{Q}_2\mathbf{Q}_1 = \mathbf{Q}_3 \tag{28}$$

and the Hermitian transpose $\tilde{\mathbf{J}}$ of the quaternion **J** is given by:

$$\tilde{\mathbf{J}} = A_0\mathbf{Q}_0 - A_1\mathbf{Q}_1 - A_2\mathbf{Q}_2 - A_3\mathbf{Q}_3 \tag{29}$$

From relations (28) and (29), as well as the properties of quaternion, it is easy to establish a complete analogy between quaternions and complex numbers.

GRAPHICAL REPRESENTATION OF POLARISATION FORMS

The Poincaré Sphere

The light ellipse may be completely defined by three independent quantities, that is, either the amplitudes A_x and A_y of the original vibration and their phase difference δ, or by the lengths of the principal semi-axes of the ellipse a and b and its azimuth ψ. Furthermore, inequalities for either of angles δ and ω define the direction of rotation of the light vector **a** describing the light ellipse. Thus:

(i)   for $0° < \delta < 180°$ or $0° < \omega < 45°$ $\qquad$ (30)

we have a right-handed elliptical polarisation, while

(ii)   for $180° < \delta < 360°$ or $-45° < \omega < 0°$ $\qquad$ (31)

we have a left-handed polarisation form.

The Poincaré sphere is a sphere of unit radius, each point of which is defined by its two spherical angular coordinates, that is the longitude and the latitude, and it represents a particular polarisation form. The longitude and the latitude of each point are taken equal to 2ψ and 2ω respectively. Figure 2 presents the Poincaré sphere of radius $R = 1$ and point P of a generic polarisation form in a right-handed Cartesian frame OXYZ. Angle 2ψ is always measured from the OX-axis in the anticlockwise sense $(0° < 2\psi \leq 360°)$ and for right-handed elliptical polarisation forms angle 2ω (positive values for ω) is taken in the upper hemisphere $(0 < 2\omega \leq 90°)$, while negative values for ω representing left-handed polarisation forms are described in the lower hemisphere $(-90° \leq 2\omega < 0)$.

In this way the following properties are valid for the Poincaré sphere:

(i)   Each point of the equator of the sphere defined by $0 \leq 2\psi < 360°$ and $\omega = 0$ represents linear polarisation forms. Furthermore the point on the positive Ox-axis represents a horizontal–linear polarisation $(\psi = 0, \omega = 0)$ (point A_{LH}), while the point on the negative OX-axis represents a linear–vertical polarisation (point A_{LV}).

(ii)   The north pole of the sphere (point N) represents a right-circular polarisation since for this point $2\omega = 90°$ and hence $a = b$. Similarly, the south pole S represents a left-circular polarisation.

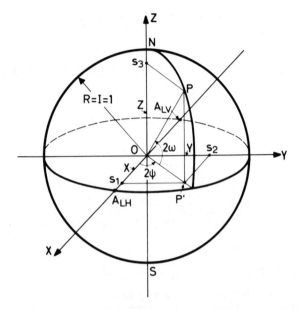

Fig. 2. Representation of an elliptically polarised light by a point P on the Poincaré sphere (longitude 2ψ and latitude 2ω of point P are equal to the double values of the azimuth ψ and the ellipticity ω of the corresponding light ellipse). The Cartesian coordinates X, Y, Z of point P represent the last three Stokes parameters of the polarised light considered.

(iii) Each point on the north hemisphere represents a right-elliptical polarisation form, while each point on the south hemisphere represents a left-elliptical polarisation form. Furthermore, the half values of the longitude 2ψ and latitude 2ω of each point of the sphere define the azimuth ψ and the ellipticity ω of the sectional pattern of the corresponding elliptical polarisation form.

(iv) Points on the same meridional ($\hat{\psi}$ = constant) represent all forms of elliptically polarised light with light ellipses of the same orientation, while points on the same parallel ($\hat{\omega}$ = constant) represent all forms with the same ellipticity.

(v) By taking the radius of Poincaré sphere R equal to unity we assume that the light intensity is equal to unity and all quantities are normalised to I. If we are interested in the absolute value of the light intensity of the elliptically polarised light, the radius of the Poincaré sphere must be taken to be equal to the light intensity (R = I).

(vi) It can be readily derived that the Cartesian coordinates X, Y, Z of a point on the Poincaré sphere are expressed by:

$$X = \cos 2\omega \cos 2\psi$$
$$Y = \cos 2\omega \sin 2\psi$$
$$Z = \sin 2\omega \qquad (32)$$

Comparing relations (32) with (17) it can be immediately derived that the Cartesian coordinates of a point on the Poincaré sphere represent the normalised to s_0 Stokes elements $s_1 s_2$ and s_3 respectively. Thus, it may be concluded that the Poincaré sphere constitutes a graphical representation of the normalised Stokes vector.

(vii) The centre O of the sphere represents the state of unpolarised light since such a light is defined as a mixture of any possible polarisation form.

It can be derived from the above-mentioned properties of the Poincaré sphere that this device yields a mapping procedure according to which each polarisation form corresponds to a point of the sphere and inversely each point of the sphere corresponds to a definite polarisation form. Thus, having in mind the sphere, one has a unified picture of all elliptical polarisation forms and it can be visualised how one form can be transformed to another by making a suitable manipulation on the sphere. Consequently one can find graphically the kind of optical device needed for transforming each polarisation form into another and inversely one can solve the problem of finding the emerging form from an optical device and a given impinging polarisation form.

However, the Poincaré sphere is a three-dimensional graphical device and although very convenient for yielding at least qualitatively short-cut solutions to various photoelastic problems it presents difficulties in the manipulation and accurate finding of the results in a three-dimensional diagram. For this reason at least two other graphical solutions were developed for this purpose which constitute various types of projections of the Poincaré sphere to different planes.

The J-circle

While in the Poincaré sphere a three-dimensional model is used to describe an elliptically polarised light, in the j-circle method the representation is a two-dimensional one. This makes it easier to represent graphically the various polarisation forms. Indeed the j-circle constitutes a parallel projection of the Poincaré sphere on its equatorial plane. The projection axis is parallel to the north–south diameter of the sphere.

In the j-circle representation of polarisation forms the following properties hold:

(i)   Each polarisation form is represented by a vector called the **j**-vector, which is the projection of the radius of the sphere connecting the point examined on the equatorial plane. The length of the **j**-vector is given by:

$$|\mathbf{j}| = \frac{a^2 - b^2}{a^2 + b^2} \tag{33}$$

where a and b are the lengths of the principal semi-axes of the light ellipse corresponding to the point examined.

(ii) The **j**-vector subtends an angle with the Ox-axis equal to the double azimuth 2ψ of the light ellipse (Fig. 3) and indicates the major axis of this light ellipse.

(iii) The tips of all **j**-vectors corresponding to various elliptical polarisation forms always lie inside the j-circle of unit radius (for cases referred to quantities normalised to the light intensity I).

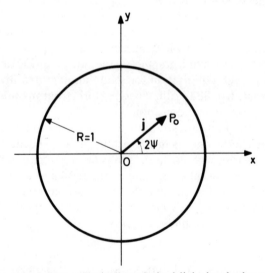

FIG. 3.   Representation of an elliptically polarised light by the **j**-vector in the j-circle method. The magnitude of the **j**-vector is equal to $|\mathbf{j}| = (a^2 - b^2)/(a^2 + b^2)$ with a and b the semi-axes of the corresponding light ellipse, while its inclination with the Ox-axis is equal to twice the azimuth ψ of the corresponding ellipse.

(iv) The tips of **j**-vectors corresponding to linear polarisation forms lie on the circumference of the j-circle with $|\mathbf{j}| = 1$ ($a \neq 0, b = 0$).

(v)   The tips of **j**-vectors for circularly polarised forms are represented by the origin of the coordinate frame $O(a = b, |\mathbf{j}| = 0)$.

The Stereographic Projection

From the various types of stereographic projections the Wulff-type net is of particular importance in polarisation optics problems. According to this type of projection a projecting pole is taken as either of the points on the sphere corresponding to intersections of the normal to the principal meridional and the equator and a projecting plane as the plane of the principal meridional. Thus, the sphere is transformed into a plane by an inversion process with a pole lying on the sphere and an inversion power equal to $2R^2$ (R = the radius of the sphere).

In the Wulff-net stereographic projection all meridional and parallel circles of the Poincaré sphere are projected as circles with the equator and the normal to the principal meridional as orthogonal diameters forming the reference frame of the projected image. This type of projection presents the main advantage to maintain the angles in the projection (conformal

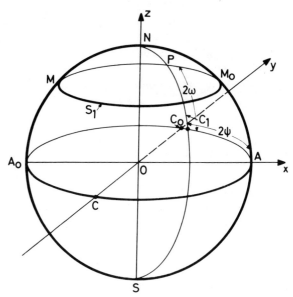

FIG. 4.   Definition of a point P on the Poincaré sphere by the meridional C_1 and parallel S_1 corresponding to the azimuth ψ and ellipticity ω.

projection). Thus, the angle between any two characteristic circles of the sphere is identical to the angle of their projections.

Let us consider a point P on the Poincaré sphere defined by the longitude 2ψ and the latitude 2ω (Fig. 4) and let us trace on the sphere equidistant meridionals (C_1), that is, great circles of constant longitude 2ψ, and

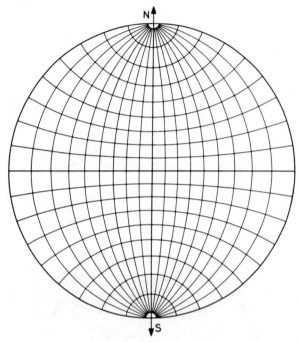

FIG. 5.   The Wulff-net used in applications with the stereographic projection of parallels and meridionals with a space equal to $10°$.

equidistant parallel circles (S_1), that is, circles of constant latitude. Their stereographic projection on the plane of the principal meridional Oxz yields a net of circular segments. The members of one family of these circles have their centres along the Ox-axis and they all pass through points N and S, while the members of the other family of circles have their centres along the Oz-axis. The net thus created by the two orthogonal families of circles on the projection plane is the so-called Wulff-net. Figure 5 presents the Wulff-net map where all meridionals and parallels were traced with a space of 10 degrees.

Thus, the created Wulff-net projection presents in detail the following properties:

(i)   The projection S_1' of each parallel S_1 of Fig. 4 is also a circle and the latitude 2ω of S_1 is equal to angle $M_0'O'A'$ subtended by the radius $O'M_0'$ to point M_0' of the intersection of S_1' with the projection of the principal meridional (outer circle of the net) with the $O'x$-axis, Fig.

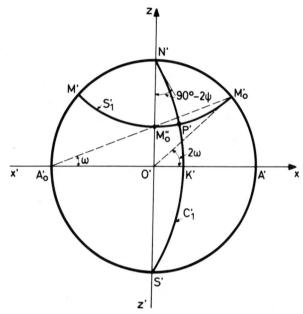

FIG. 6.   Determination of a point P' in the Wulff-net by the stereographic projection of the meridional C_1 and parallel S_1 of Fig. 4 to the circles C_1' and S_1' in the net.

6. For the complete determination of circle S_1' a third point M_0'' must be defined at the intersection of $A_0'M_0'$ with the $O'z$-axis, where line $A_0'M'$ subtends an angle ω with the $x'O'x$-axis.

(ii) The projection of the equator of the Poincaré sphere is represented by the $x'O'x$-axis, the projection of the normal to the principal meridional by $z'O'z$-axis and the principal meridional as the outer circle of the net.

(iii) The projection C_1' of each meridional C_1 of Fig. 4 is also a circle passing through points N' and S' of the net. A third point K' of each

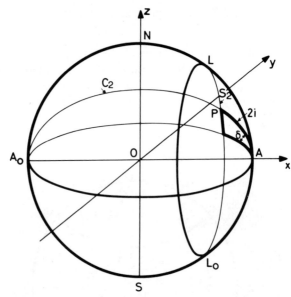

FIG. 7. Definition of a point P on the Poincaré sphere by the parallel to the principal meridional S_2 and the major circle through the Ox-axis C_2 corresponding to the amplitude ratio i and the phase retardation δ.

circle is defined at the intersection of this circle with the equator ($x'O'x$-axis). Point K′ is defined on the $x'O'x$-axis by transferring the subdivisions of the diameter $z'O'z$ intersected by the circle S_1' to corresponding subdivisions of the diameter $x'O'x$. Thus, if the longitude of meridional NPS of Fig. 4 is equal to 2ψ the length O′K′ is equal to the length defined on the $z'O'z$-axis by the circle S_1' with $2\omega = 2\psi$.

(iv) The angles subtended between characteristic circles of the Poincaré sphere are maintained in the projection. Thus, the projections of parallels and meridionals constitute orthogonal pairs of curves. This property is very useful in defining the tangents of circle C_1' at points N′ and S′ which both subtend angles equal to $90° - 2\psi$ with the $z'O'z$-axis.

(v)   In the case when the position of point P on the Poincaré sphere is defined by angles δ subtended between the great circle C_2 and the equator and i defined by the arc $\widehat{AP}$ of the circle C_2 of Fig. 7, we trace on the sphere the corresponding circles C_2 and S_2 through the Ox-axis and normal to the principal meridional respectively and

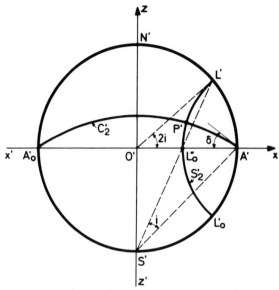

FIG. 8. Determination of a point P' in the Wulff-net by the stereographic projection of the circles C_2 and S_2 of Fig. 7 to the circles C'_2 and S'_2 in the net.

finally project this system of circles stereographically on the plane of principal meridional. We thus obtain a new Wulff-net similar to the previous one, but angularly displaced to the first net by $90°$. Figure 8 indicates the properties of the new Wulff-net. Both orthogonal nets are traced in Fig. 5 and both may be used for the complete and accurate evaluation of the angular coordinates of the emerging polarised light from an optical element.

Interrelation Between Analytical and Graphical Methods for the Characterisation of an Elliptically Polarised Light

The above-established analytic and graphical methods of characterisation of a polarisation form were introduced independently to each other in terms of the parameters necessary to define an elliptical polarisation form. However, these methods are not independent of each other and each one of them may be expressed in terms of the others. Now we shall establish rapidly the correlations existing between these methods.

Since both the Stokes and Jones vectors may describe the same polarisation form there must exist a relationship between these two vectors.

If **a** represents the Jones vector for a given polarisation form which is also represented by the corresponding Stokes vector **S,** vectors **S** and **a** are expressed by relations (15) and (21) respectively. It can be readily found that the following relations hold between Stokes parameters s_i and Jones vector **a** and its Hermitian conjugate **ã**:

$$s_0 = \tilde{\mathbf{a}}\mathbf{Q}_0\mathbf{a}$$

$$s_1 = \tilde{\mathbf{a}}\mathbf{Q}_1\mathbf{a}$$

$$s_2 = \tilde{\mathbf{a}}\mathbf{Q}_2\mathbf{a}$$

$$s_3 = \tilde{\mathbf{a}}\mathbf{Q}_3\mathbf{a} \tag{34}$$

where the quantities $\mathbf{Q}_i$ $(i = 1, 2, 3, 4)$ are given by the Pauli matrices expressed by relations (26).

As for the interrelation between the Stokes vector and the Poincaré sphere representation we have already indicated that, while Stokes parameter s_0 corresponds to the intensity of the impinging light, the remaining Stokes parameters s_1 s_2 and s_3 correspond to the Cartesian coordinates of the corresponding point P in the Poincaré sphere.

In order to show that the parallel projection of point P on the Poincaré sphere on its equatorial plane, which yields point P' (Fig. 2), defines the corresponding **j**-vector OP' of the same polarisation form, we express the projection OP' in terms of angle ω and by introducing from relation (13) the expression for $\tan\omega = \pm b/a$ we obtain:

$$OP' = \cos 2\omega = \frac{1 - \tan^2\omega}{1 + \tan^2\omega} = \frac{a^2 - b^2}{a^2 + b^2}$$

which indicates that the vector OP' coincides with the respective **j**-vector.

Although we are interested in this chapter only with polarised light it is worthwhile mentioning here that the Stokes vector is convenient not only for polarised and partially polarised light, but also it can be used for totally unpolarised light. Indeed, the Stokes parameters were first introduced as time-average quantities and their validity was proved based on electromagnetic theory of light. For the case of unpolarised light which does not present time-average preference between the components A_x and A_y and δ, the Stokes vector can be presented as:

$$\mathbf{S} = \begin{bmatrix} (A_x^2 + A_y^2) \\ (A_x^2 - A_y^2) \\ (2A_xA_y\cos\delta) \\ (2A_xA_y\sin\delta) \end{bmatrix} = \begin{bmatrix} (2A_x^2) \\ 0 \\ 0 \\ 0 \end{bmatrix} = (2A_x^2)\begin{bmatrix} 1 \\ 0 \\ 0 \\ 0 \end{bmatrix}$$

where angle-brackets denote time averages of the various quantities.

Contrary to Stokes vector, the Jones vector is convenient only for polarised light, which is described by this two-element complex column vector with high simplicity and elegance. In this respect the Stokes vector is particularly relevant in the incoherent addition of two light beams, while the Jones vector is especially suitable for determining associations of coherent polarised-light beams.

Complementary to Stokes and Jones vectors, the Poincaré sphere and the other graphical methods derived from the sphere by projection, which are suitable for describing indifferently totally or partially polarised light, are especially convenient for predicting graphically the same transformations.

Thus, all these methods, which are characterised by an extreme elegance and compactness, provide in a unified way a complete definition of all kinds of polarisation regardless of being elliptical, circular or linear by yielding the azimuth, the ellipticity and the handedness of the light ellipse for the light emerging from an optical element, which is illuminated by some kind of polarised light. However, the main advantage of all these methods is not related to a uniform representation of each separate polarisation form, but mainly to the fact that they provide a direct solution to the problem of finding the emerging polarisation form from an optical device when the impinging polarisation form to it is known.

EFFECT OF AN OPTICAL ELEMENT TO POLARISED LIGHT

Definition of Optical Elements

Before proceeding to the prediction of the effect of an optical element on a polarised light beam we shall give briefly the definition of the most common optical elements used in the analysis.

Two forms of elliptical polarisation are called orthogonal, if their corresponding light ellipses have the same ellipticity, their azimuths differ by 90° and they have opposite handedness.

A polariser is an optical element which divides the incident light beam into two orthogonal forms and transmits either form with a different degree of transmittance. The action of a polariser is described by its two eigenvectors and its two transmission coefficients. An eigenvector is defined as a special polarisation form which emerges unchanged from the polariser. An eigenvector of the polariser can be depicted by a point on the Poincaré sphere, or by a Stokes or a Jones vector. The transmittance coefficient of a polariser along an eigenvector is defined as the ratio of the intensities of the transmitted to the incident light beams, these two beams being described by the eigenvector of the polariser. The transmittance coefficients are designated by k_1 and k_2 and for the polarisers used here we assume that $k_1 = 1$ and $k_2 = 0$. If the polarisation formed by a polariser is linear, circular or elliptical, the polariser is called linear, circular or elliptical polariser respectively. Thus, for example, a linear polariser whose eigenvector corresponds to a horizontal linear polarisation transmits without any loss of intensity the totality of the horizontally polarised light and extinguishes all the vertically polarised light.

A retarder or birefringent plate is an optical element that divides an incident monochromatic polarised light beam into two orthogonal polarisation forms and yields to either form a certain optical retardation relative to the other. The two polarisation forms which are conserved by the retarder are called the eigenvectors of the retarder and whether they are linear, circular or elliptical, the retarder is called linear, circular or elliptical respectively. The eigenvector corresponding to the smaller refractive index of the retarder is called the fast eigenvector and the eigenvector corresponding to the higher refractive index is called the slow eigenvector. A retarder is characterised by its own eigenvectors and by its optical retardation.

A circular retarder is frequently referred to as an optically active device or a rotator, while a linear retarder with a retardation equal to $90°$ or $180°$ is called a quarter-wave plate or a half-wave plate respectively.

Transformation of Polarised Light Passing Through Either a Polariser or a Retarder

We establish now the transformation of a polarised light beam when it passes through either a polariser or a retarder. The cases of a linear polariser and a linear retarder will be treated in detail, since these are the most common devices used in photoelasticity. However, for purposes of completeness we shall describe also the influence of the more general cases of an elliptical polariser and an elliptical retarder on a monochromatic light beam.

According to Stokes-vector method each particular light form is fully characterised by four real quantities, the Stokes elements, s_i ($i = 0, 1, 2, 3$), which constitute the components of the Stokes four-element column vector.

The use of Stokes vector for the solution of polarisation optics problems was introduced by Mueller.[2] Mueller has established his calculus based on the fact that, if the four Stokes parameters are considered as the elements of a four-element column vector, then since the incident and emerging light beams to and from the optical element are presented by such vectors, the element itself must be represented by a four-by-four real matrix.

The complete characterisation of an optical element by such a matrix leads to an elegant and unified determination of the emerging light from the device by knowing the characteristics of the incident light beam. For this purpose it is only necessary to multiply the column vector of the incident light by the square matrix of the optical element and the result is the Stokes vector for the emerging light. It is self-evident that the matrix multiplication process can be applied to any number of optical elements interposed in the light path.

The Jones calculus is based on the corresponding Jones vector characterising an elliptical polarisation form in the same way as the Mueller calculus is based on Stokes vector. However, Jones calculus can only be applied to polarised light forms whereas Mueller calculus finds an application for any form of light.

Since any elliptically polarised light is represented by a two-element complex column vector, the matrix operator which connects two such vectors for the incident and the emerging light form must be a two-by-two complex matrix. Thus, any optical element must be represented according to Jones calculus by a two-by-two complex matrix. This matrix is called the Jones matrix of the optical element and the corresponding calculus the Jones calculus.

THE EFFECT OF A POLARISER OR A RETARDER ON A POLARISED LIGHT BEAM

We establish now what happens to a polarised light beam when it passes through either a polariser or a retarder. The cases of linear polarisers and retarders will be treated in detail since these are the most common optical elements used in photoelasticity. However, for purposes of completeness, the influence of the more general cases of elliptical polarisers and retarders on a monochromatic light beam will be studied also.

The Jones Calculus for the Linear Polariser and Retarder
Let us consider the case of a linear polariser whose eigenvector is a linearly polarised light subtending an angle θ with the Ox-axis, that is a linearly polarised light at an angle θ with the Ox-axis passes through the polariser without any change. Suppose also that a linearly polarised light of

amplitude A at an angle φ with the Ox-axis is incident to the polariser. The light vector $\mathbf{a}$ representing the impinging light can be expressed as:

$$\mathbf{a} = a_x\mathbf{i} + a_y\mathbf{j} = A\cos\varphi\mathbf{i} + A\sin\varphi\mathbf{j}$$

where $\mathbf{i}$ and $\mathbf{j}$ are the unit vectors along the reference frame Oxy, normal to the direction of propagation of light (Oz-axis).

We define the incident and the emergent light beams by the Jones vectors:

$$\begin{bmatrix} a_x \\ a_y \end{bmatrix} = \begin{bmatrix} A_x\exp(i\delta_x) \\ A_y\exp(i\delta_y) \end{bmatrix} \quad \text{and} \quad \begin{bmatrix} a'_x \\ a'_y \end{bmatrix} = \begin{bmatrix} A'_x\exp(i\delta'_x) \\ A'_y\exp(i\delta'_y) \end{bmatrix}$$

The polariser transmits only the component of the $\mathbf{a}$-vector along its axis at angle θ with the Ox-direction. If the amplitude of this component is analysed along the Ox- and Oy-axes the components of the emerging-light vector $\mathbf{a}'$ from the polariser will be obtained. Thus, we have the relation:

$$\begin{vmatrix} A'_x\exp(i\delta'_x) \\ A'_y\exp(i\delta'_y) \end{vmatrix} = \begin{vmatrix} \cos^2\theta & \sin\theta\cos\theta \\ \sin\theta\cos\theta & \sin^2\theta \end{vmatrix}\begin{vmatrix} A_x\exp(i\delta_x) \\ A_y\exp(i\delta_y) \end{vmatrix} \tag{35}$$

For the linear retarder we have the light vector $\mathbf{a}$ impinging a retarder with its fast axis arranged at an angle θ with the Ox-axis and whose retardation is equal to δ. The components a_f and a_s of $\mathbf{a}$ along the fast and slow axes of the retarder at the entrance are given by:

$$a_f = a_x\cos\theta + a_y\sin\theta$$
$$a_s = -a_x\sin\theta + a_y\cos\theta$$

At the exit of the retarder the a_s-component is retarded by δ, relative to the a_f-component, so that at the exit of the retarder the light components along the fast (a'_f) and slow (a'_s) axes are given by:

$$a'_f = a_f\exp(i\delta) \quad a'_s = a_s$$

Then the components a'_x and a'_y emerging from the linear retarder light beam along the Ox- and Oy-axes are given by:

$$\begin{bmatrix} A'_x\exp(i\delta'_x) \\ A'_y\exp(i\delta'_y) \end{bmatrix} = \begin{bmatrix} \exp(i\delta)\cos^2\theta + \sin^2\theta & [\exp(i\delta) - 1)]\sin\theta\cos\theta \\ [\exp(i\delta) - 1)]\sin\theta\cos\theta & \exp(i\delta)\sin^2\theta + \cos^2\theta \end{bmatrix}$$

$$\times \begin{bmatrix} A_x\exp(i\delta_x) \\ A_y\exp(i\delta_y) \end{bmatrix} \tag{36}$$

The above two-by-two matrices included in relations (35) and (36) which characterise an optical element (polariser or retarder) are called Jones matrices and the corresponding calculus is called Jones calculus. By knowing the Jones matrix of an optical element we can calculate the Jones vector of the emergent light beam, provided we know the Jones vector of the incident light, by multiplying the Jones matrix of the device by the Jones vector of the incident light.

Thus, if we denote by $\mathbf{a}$, $\mathbf{a}'$ and $\mathbf{J}$ the Jones vectors of the incident and emerging light beams and the Jones matrix of the optical element respectively, we have:

$$\mathbf{a}' = \mathbf{J}\mathbf{a} \tag{37}$$

Relation (37) constitutes the key formula of the Jones calculus. Thus, if we have two optical elements in series with Jones matrices $\mathbf{J}_1$ and $\mathbf{J}_2$ and the Jones vector $\mathbf{a}_1$ of the original incident light beam, we can find the Jones vectors of the emerging light from the first element $\mathbf{a}_2$ and the second element $\mathbf{a}_3$ respectively, since it is valid that:

$$\mathbf{a}_2 = \mathbf{J}_1\mathbf{a}_1 \quad \text{and} \quad \mathbf{a}_3 = \mathbf{J}_2\mathbf{a}_2$$

and therefore:

$$\mathbf{a}_3 = \mathbf{J}_2(\mathbf{J}_1\mathbf{a}_1) = \mathbf{J}_2\mathbf{J}_1\mathbf{a}_1 \tag{38}$$

Similarly, if we consider a train of n-optical elements with Jones matrices $\mathbf{J}_1, \mathbf{J}_2, \mathbf{J}_3, \ldots, \mathbf{J}_n$ respectively and a light beam with Jones vector $\mathbf{a}_0$, which enters successively all the elements in the order prescribed, the Jones vector of the emerging light is expressed by:

$$\mathbf{a} = \mathbf{J}_n\mathbf{J}_{(n-1)} \cdots \mathbf{J}_2\mathbf{J}_1\mathbf{a}_0 \tag{39}$$

It can readily be shown that:

$$\begin{bmatrix} \cos^2\theta & \sin\theta\cos\theta \\ \sin\theta\cos\theta & \sin^2\theta \end{bmatrix} = \begin{bmatrix} \cos\theta & -\sin\theta \\ \sin\theta & \cos\theta \end{bmatrix}\begin{bmatrix} 1 & 0 \\ 0 & 0 \end{bmatrix}\begin{bmatrix} \cos\theta & \sin\theta \\ -\sin\theta & \cos\theta \end{bmatrix} \tag{40}$$

where the matrix $\mathbf{J}_0 = \begin{bmatrix} 1 & 0 \\ 0 & 0 \end{bmatrix}$ represents the Jones matrix of a polariser with a horizontal optical axis ($\theta = 0$). Thus, relation (40) may be written as:

$$\mathbf{J}_\theta = \mathbf{R}(-\theta)\mathbf{J}_0\mathbf{R}(\theta) \tag{41}$$

where $\mathbf{J}_\theta$ is the Jones matrix of the polariser with an eigenvector at angle θ with the Ox-axis, $\mathbf{J}_0$ is the corresponding matrix with $\theta = 0$ and $\mathbf{R}(\theta)$ is the rotation matrix given by:

$$\mathbf{R}(\theta) = \begin{bmatrix} \cos\theta & \sin\theta \\ -\sin\theta & \cos\theta \end{bmatrix} \tag{42}$$

Similarly it can be seen that:

$$\begin{bmatrix} \exp(i\delta)\cos^2\theta + \sin^2\theta & [\exp(i\delta) - 1)]\sin\theta\cos\theta \\ [\exp(i\delta) - 1)]\sin\theta\cos\theta & \exp(i\delta)\sin^2\theta + \cos^2\theta \end{bmatrix} = \begin{bmatrix} \cos\theta & -\sin\theta \\ \sin\theta & \cos\theta \end{bmatrix}$$

$$\times \begin{bmatrix} \exp(i\delta) & 0 \\ 0 & 1 \end{bmatrix} \begin{bmatrix} \cos\theta & \sin\theta \\ -\sin\theta & \cos\theta \end{bmatrix} \quad (43)$$

where the matrix $\mathbf{J}_0 = \begin{bmatrix} \exp(i\delta) & 0 \\ 0 & 1 \end{bmatrix}$ represents the matrix of the linear retarder with its fast axis coinciding with the Ox-axis. Thus, relation (43) can be put in the form (41) and therefore it is valid that:

$$\mathbf{J}(\theta + \varphi) = \mathbf{R}(-\theta)\mathbf{J}(\varphi)\mathbf{R}(\theta) \quad (44)$$

where $\mathbf{J}(\theta + \varphi)$ represents the Jones matrix of an optical element with eigenvector at an angle $(\theta + \varphi)$ and $\mathbf{J}(\varphi)$ the relevant matrix of the same element with eigenvector at angle φ.

Relation (44) constitutes the basic formula for transforming the Jones matrix of an optical element at a given orientation φ into the Jones matrix of the same element at the new orientation $(\theta + \varphi)$.

The Jones matrices for the common types of polarisers and retarders are listed in Tables 3 and 4 respectively.

In order to define the Jones matrices for the more general case of an elliptical polariser and retarder we use the same procedure as previously, by also applying the linearity principle. The expressions for these matrices are given in Tables 3 and 4 respectively.

The Mueller Calculus for the Linear Polariser and Retarder

The Mueller calculus is associated with the Stokes vector, which is described by a four-element column vector. Thus, since the incident and the emerging light beams from an optical element are described by four element column vectors it is self-evident that the optical device should be described by a four-by-four real matrix. This matrix is called the Mueller matrix and the calculus based on Mueller matrices and Stokes vectors is called the Mueller calculus.

The Mueller matrix of an optical element can be found by using the linearity principle and the special characteristic properties of the optical element considered. This principle can be expressed in the form:

$$\begin{bmatrix} s_0' \\ s_1' \\ s_2' \\ s_3' \end{bmatrix} = \begin{bmatrix} m_{11} & m_{12} & m_{13} & m_{14} \\ m_{21} & m_{22} & m_{23} & m_{24} \\ m_{31} & m_{32} & m_{33} & m_{34} \\ m_{41} & m_{42} & m_{43} & m_{44} \end{bmatrix} \begin{bmatrix} s_0 \\ s_1 \\ s_2 \\ s_3 \end{bmatrix} \quad (45)$$

TABLE 3
JONES AND MUELLER MATRICES FOR POLARISERS

Type of polariser	Jones matrix	Mueller matrix
Ideal Elliptical, producing elliptically polarised light with azimuth ψ and ellipticity ω such that: $\tan 2\psi = \tan 2\theta \cos \delta$; $\sin 2\omega = \sin 2\theta \sin \delta$	$\begin{bmatrix} \cos^2\theta & e^{-i\delta}\sin\theta\cos\theta \\ e^{i\delta}\sin\theta\cos\theta & \sin^2\theta \end{bmatrix}$	$\frac{1}{2}\begin{bmatrix} 1 & \cos 2\theta & \sin 2\theta\cos\delta & \sin 2\theta\sin\delta \\ \cos 2\theta & \cos^2 2\theta & \sin 2\theta\cos 2\theta\cos\delta & \sin 2\theta\cos 2\theta\sin\delta \\ \sin 2\theta\cos\delta & \sin 2\theta\cos 2\theta\cos\delta & \sin^2 2\theta\cos^2\delta & \sin^2 2\theta\sin\delta\cos\delta \\ \sin 2\theta\sin\delta & \sin 2\theta\cos 2\theta\sin\delta & \sin^2 2\theta\sin\delta\cos\delta & \sin^2 2\theta\sin^2\delta \end{bmatrix}$
Ideal Right Circular, ($\theta = 45°$, $\delta = 90°$)	$\frac{1}{2}\begin{bmatrix} 1 & -i \\ i & 1 \end{bmatrix}$	$\frac{1}{2}\begin{bmatrix} 1 & 0 & 0 & 1 \\ 0 & 0 & 0 & 0 \\ 0 & 0 & 0 & 0 \\ 1 & 0 & 0 & 1 \end{bmatrix}$
Ideal Left Circular, ($\theta = 45°$, $\delta = -90°$)	$\frac{1}{2}\begin{bmatrix} 1 & i \\ -i & 1 \end{bmatrix}$	$\frac{1}{2}\begin{bmatrix} 1 & 0 & 0 & -1 \\ 0 & 0 & 0 & 0 \\ 0 & 0 & 0 & 0 \\ -1 & 0 & 0 & 1 \end{bmatrix}$
Ideal Linear, with its axis at angle θ ($\delta = 0$)	$\begin{bmatrix} \cos^2\theta & \sin\theta\cos\theta \\ \sin\theta\cos\theta & \sin^2\theta \end{bmatrix}$	$\frac{1}{2}\begin{bmatrix} 1 & \cos 2\theta & \sin 2\theta & 0 \\ \cos 2\theta & \cos^2 2\theta & \sin 2\theta\cos 2\theta & 0 \\ \sin 2\theta & \sin 2\theta\cos 2\theta & \sin^2 2\theta & 0 \\ 0 & 0 & 0 & 0 \end{bmatrix}$
Partial Linear, with principal intensity transmission coefficients k_1 and k_2 along its principal directions 1 and 2	$\begin{bmatrix} \sqrt{k_1} & 0 \\ 0 & \sqrt{k_2} \end{bmatrix}$	$\frac{1}{2}\begin{bmatrix} k_1+k_2 & k_1-k_2 & 0 & 0 \\ k_1-k_2 & k_1+k_2 & 0 & 0 \\ 0 & 0 & 2\sqrt{k_1 k_2} & 0 \\ 0 & 0 & 0 & 2\sqrt{k_1 k_2} \end{bmatrix}$

TABLE 4
JONES AND MUELLER MATRICES FOR RETARDERS

Type of retarder	Jones matrix	Mueller matrix
Elliptical, giving two orthogonal elliptical polarisation forms with azimuth ψ, ellipticity ω and retardation δ.	$\begin{bmatrix} \exp(i\delta)\cos^2\theta + \sin^2\theta & [\exp(i\delta)-1]\exp(-i\Delta)\sin\theta\cos\theta \\ [\exp(i\delta)-1]\exp(i\Delta)\sin\theta\cos\theta & \cos^2\theta + \exp(i\delta)\sin^2\theta \end{bmatrix}$ with: $A_1 = \cos 2\omega\cos 2\psi\sin\frac{\delta}{2}$ $A_3 = \sin 2\omega\sin\frac{\delta}{2}$	$\begin{bmatrix} 1 & 0 & 0 & 0 \\ 0 & A_1^2 - A_2^2 - A_3^2 + A_4^2 & 2(A_1 A_2 + A_3 A_4) & -2(A_1 A_3 + A_2 A_4) \\ 0 & 2(A_1 A_2 - A_3 A_4) & -A_1^2 + A_2^2 - A_3^2 + A_4^2 & 2(A_1 A_4 - A_2 A_3) \\ 0 & -2(A_1 A_3 - A_2 A_4) & -2(A_1 A_4 + A_2 A_3) & -A_1^2 - A_2^2 + A_3^2 + A_4^2 \end{bmatrix}$ $A_2 = \cos 2\omega\sin 2\psi\sin\frac{\delta}{2}$ $A_4 = \cos\frac{\delta}{2}$
Right and left circular, ($\theta = 45°$, $\Delta = \pm 90°$) Upper signs for right, lower signs for left circular	$\begin{bmatrix} \cos\frac{\delta}{2} & \pm\sin\frac{\delta}{2} \\ \mp\sin\frac{\delta}{2} & \cos\frac{\delta}{2} \end{bmatrix}$	$\begin{bmatrix} 1 & 0 & 0 & 0 \\ 0 & \cos\delta & \pm\sin\delta & 0 \\ 0 & \mp\sin\delta & \cos\delta & 0 \\ 0 & 0 & 0 & 1 \end{bmatrix}$
Linear, with its fast axis at an angle θ and retardation δ ($\Delta = 0$, $\psi = \theta$)	$\begin{bmatrix} \exp(i\delta)\cos^2\theta + \sin^2\theta & [\exp(i\delta)-1]\sin\theta\cos\theta \\ [\exp(i\delta)-1]\sin\theta\cos\theta & \exp(i\delta)\sin^2\theta + \cos^2\theta \end{bmatrix}$	$\begin{bmatrix} 1 & 0 & 0 & 0 \\ 0 & \cos^2 2\theta + \sin^2 2\theta\cos\delta & (1-\cos\delta)\sin 2\theta\cos 2\theta & -\sin 2\theta\sin\delta \\ 0 & (1-\cos\delta)\sin 2\theta\cos 2\theta & \sin^2 2\theta + \cos^2 2\theta\cos\delta & \cos 2\theta\sin\delta \\ 0 & \sin 2\theta\sin\delta & -\cos 2\theta\sin\delta & \cos\delta \end{bmatrix}$
Half-wave plate, with its fast axis at an angle θ ($\delta = 180°$)	$\begin{bmatrix} -\cos 2\theta & -\sin 2\theta \\ -\sin 2\theta & \cos 2\theta \end{bmatrix}$	$\begin{bmatrix} 1 & 0 & 0 & 0 \\ 0 & \cos 4\theta & \sin 4\theta & 0 \\ 0 & \sin 4\theta & -\cos 4\theta & 0 \\ 0 & 0 & 0 & -1 \end{bmatrix}$
Quarter-wave plate, with its fast axis at an angle θ ($\delta = 90°$)	$\begin{bmatrix} i\cos^2\theta + \sin^2\theta & (i-1)\sin\theta\cos\theta \\ (i-1)\cos\theta\sin\theta & i\sin^2\theta + \cos^2\theta \end{bmatrix}$	$\begin{bmatrix} 1 & 0 & 0 & 0 \\ 0 & \cos^2 2\theta & \sin 2\theta\cos 2\theta & -\sin 2\theta \\ 0 & \sin 2\theta\cos 2\theta & \sin^2 2\theta & \cos 2\theta \\ 0 & \sin 2\theta & -\cos 2\theta & 0 \end{bmatrix}$

where the column matrix with primes in the left-hand side of relation (45) represents the Stokes vector of the emerging light beam, the column matrix in the right-hand side without primes represents the incident light beam and the four-by-four matrix the Mueller matrix of the optical element. Relation (45) may be written in a condensed form as:

$$\mathbf{S'} = \mathbf{MS} \tag{46}$$

Let us now consider a linear polariser whose eigenvector subtends an angle θ with the Ox axis. We can calculate the coefficients m_{ij} $(i, j = 1, 2, 3, 4)$ of the matrix $\mathbf{M}$ by using the following properties of the linear polariser:

(i) An unpolarised light beam passing through a linear polariser becomes linearly polarised at an angle θ with the Ox-axis. That is:

$$\text{If:} \quad \mathbf{S} = \begin{bmatrix} 2 \\ 0 \\ 0 \\ 0 \end{bmatrix} \quad \text{then:} \quad \mathbf{S'} = \begin{bmatrix} 1 \\ \cos 2\theta \\ \sin 2\theta \\ 0 \end{bmatrix}$$

(ii) A linearly polarised light at an angle θ with the Ox-axis remains unaffected that is:

$$\text{If:} \quad \mathbf{S} = \begin{bmatrix} 1 \\ \cos 2\theta \\ \sin 2\theta \\ 0 \end{bmatrix} \quad \text{then:} \quad \mathbf{S'} = \begin{bmatrix} 1 \\ \cos 2\theta \\ \sin 2\theta \\ 0 \end{bmatrix}$$

(iii) A linearly polarised light along the Ox-axis becomes linearly polarised at an angle θ with the Ox-axis. If the intensity of the incident light is taken equal to unity then the Stokes parameters of the emerging light take the values given by $\mathbf{S'}$, that is:

$$\text{If:} \quad \mathbf{S} = \begin{bmatrix} 1 \\ 1 \\ 0 \\ 0 \end{bmatrix} \quad \text{then:} \quad \mathbf{S'} = \begin{bmatrix} \cos^2 \theta \\ \cos^2 \theta \cos 2\theta \\ \cos^2 \theta \sin 2\theta \\ 0 \end{bmatrix}$$

(iv) A right-circularly polarised light of unit intensity passing through the linear polariser becomes linearly polarised at an angle θ with the

Ox-axis with an intensity equal to half the intensity of the incident light, that is:

$$\text{If:} \quad \mathbf{S} = \begin{bmatrix} 1 \\ 0 \\ 0 \\ 1 \end{bmatrix} \quad \text{then:} \quad \mathbf{S}' = \frac{1}{2} \begin{bmatrix} 1 \\ \cos 2\theta \\ \sin 2\theta \\ 0 \end{bmatrix}$$

By introducing each of the pairs of values for $\mathbf{S}$ and $\mathbf{S}'$ given by the four conditions above-mentioned we deduce the values of the coefficients m_{ij} of the Mueller matrix $\mathbf{M}$, which for the particular case of a linear polariser is designated as:

$$\mathbf{P}_\theta = \frac{1}{2} \begin{bmatrix} 1 & \cos 2\theta & \sin 2\theta & 0 \\ \cos 2\theta & \cos^2 2\theta & \sin 2\theta \cos 2\theta & 0 \\ \sin 2\theta & \sin 2\theta \cos 2\theta & \sin^2 2\theta & 0 \\ 0 & 0 & 0 & 0 \end{bmatrix} \tag{47}$$

The Mueller matrices for the commonly used polarisers in the praxis of polarisation optics are tabulated in Table 3.

We shall now try to establish the Mueller matrix of a linear retarder with a relative retardation δ, whose fast axis subtends an angle θ with the Ox-axis by applying the linearity principle expressed by relation (45) and the basic properties of a retarder. These properties are expressed as follows:

(i) An unpolarised light beam passing through a linear retarder remains unaffected, that is:

$$\text{If:} \quad \mathbf{S} = \begin{bmatrix} 1 \\ 0 \\ 0 \\ 0 \end{bmatrix} \quad \text{then:} \quad \mathbf{S}' = \begin{bmatrix} 1 \\ 0 \\ 0 \\ 0 \end{bmatrix}$$

(ii) A linearly polarised light beam passing through the fast axis of a linear retarder remains unaffected, that is:

$$\text{If:} \quad \mathbf{S} = \begin{bmatrix} 1 \\ \cos 2\theta \\ \sin 2\theta \\ 0 \end{bmatrix} \quad \text{then:} \quad \mathbf{S}' = \begin{bmatrix} 1 \\ \cos 2\theta \\ \sin 2\theta \\ 0 \end{bmatrix}$$

(iii) The components a_x, a_y of a linearly polarised light of unit amplitude along the Ox-axis with $a_x = \exp(i\delta_x)$, $a_y = 0$ after passing through the linear retarder, become:

$$a'_x = [\exp(i\delta)\cos^2\theta + \sin^2\theta]a_x \quad a'_y = \exp[(i\delta) - 1]\sin\theta\cos\theta\, a_x$$

Then:

If: $\quad \mathbf{S} = \begin{bmatrix} 1 \\ 1 \\ 0 \\ 0 \end{bmatrix} \quad$ then: $\quad \mathbf{S}' = \begin{bmatrix} 1 \\ \cos^2 2\theta + \sin^2 2\theta\cos\delta \\ (1 - \cos\delta)\sin 2\theta\cos 2\theta \\ \sin 2\theta\sin\delta \end{bmatrix}$

(iv) Similarly, a right-circularly polarised light beam $\mathbf{S}$ passing through a linear retarder becomes a light beam expressed by the following matrix $\mathbf{S}'$, that is:

If: $\quad \mathbf{S} = \begin{bmatrix} 1 \\ 0 \\ 0 \\ 1 \end{bmatrix} \quad$ then: $\quad \mathbf{S}' = \begin{bmatrix} 1 \\ -\sin 2\theta\sin\delta \\ \cos\theta\sin\delta \\ \cos\delta \end{bmatrix}$

Again by introducing each of the pairs of values for $\mathbf{S}$ and $\mathbf{S}'$ derived from the four conditions above-mentioned we deduce the values of the coefficients m_{ij} of the Mueller matrix $\mathbf{M}$ which for the case of the linear retarder is designated by $\mathbf{R}_\theta$ and given by the expression:

$$\mathbf{R}_\theta = \begin{bmatrix} 1 & 0 & 0 & 0 \\ 0 & \cos^2 2\theta + \sin^2 2\theta\cos\delta & (1 - \cos\delta)\sin 2\theta\cos 2\theta & -\sin 2\theta\sin\delta \\ 0 & (1 - \cos\delta)\sin 2\theta\cos 2\theta & \sin^2 2\theta + \cos^2 2\theta\cos\delta & \cos 2\theta\sin\delta \\ 0 & \sin 2\theta\sin\delta & \cos 2\theta\sin\delta & \cos\delta \end{bmatrix} \quad (48)$$

The Mueller matrices for the more general cases of elliptical polarisers and retarders can be found in an analogous manner. The Mueller matrices for the commonly used retarders in polarisation optics are listed in Table 4.

As in the case of Jones calculus it can also be proved for the case of Mueller calculus that if we have a train of n optical devices with Mueller matrices $\mathbf{M}_n$ respectively and consider that a light beam with Stokes vector $\mathbf{S}_0$ impinges the first optical element and subsequently all the others in series, the polarised light beam emerging from the last optical element (n) is given by:

$$\mathbf{S} = \mathbf{M}_n\mathbf{M}_{n-1}\cdots\mathbf{M}_2\mathbf{M}_1\mathbf{S}_0 \qquad (49)$$

It can also be proved, as in the case of Jones calculus, that the Mueller matrix of an optical element whose principal axis subtends an angle $(\theta + \varphi)$ with the Ox-axis is related with the Mueller matrix of the same optical element with its principal axis subtending an angle φ with the Ox-axis through the relation:

$$\mathbf{M}(\theta + \varphi) = \mathbf{T}(-2\theta)\mathbf{M}(\varphi)\mathbf{T}(2\theta) \qquad (50)$$

where $\mathbf{T}(2\theta)$ is the well known rotator matrix given by:

$$\mathbf{T}(2\theta) = \begin{bmatrix} 1 & 0 & 0 & 0 \\ 0 & \cos 2\theta & \sin 2\theta & 0 \\ 0 & -\sin 2\theta & \cos 2\theta & 0 \\ 0 & 0 & 0 & 1 \end{bmatrix} \qquad (51)$$

The Poincaré Sphere for the Linear Polariser and Retarder

It has been already shown that the Poincaré sphere may be considered as the geometrical representation of the Stokes vector. Then, the transformations on the Poincaré sphere necessary for finding the effect of either a polariser or a retarder on an elliptically polarised light beam can be readily established from the corresponding transformations of the Stokes vector by using the Mueller calculus. By transforming the Stokes elements s_1, s_2, s_3 for the incident and the emerging light beam into the Cartesian coordinates of points on the Poincaré sphere we obtain on the sphere two points representing the incident and the emerging polarisation forms. However, the convenient transformations on the Poincaré sphere for finding the effect of an optical element on a polarised light beam may be also established independently of the above-mentioned procedure.

For this purpose we consider first a linear polariser with its eigenvector subtending an angle θ with the Ox-axis and an elliptically polarised light beam of unit intensity which is incident on the polariser with azimuth ψ and ellipticity ω ($\tan \omega = b/a$). Then the intensity of the emerging light beam is given by:

$$I = \tfrac{1}{2} + \tfrac{1}{2}\cos 2\omega \cos 2(\psi - \theta) \qquad (52)$$

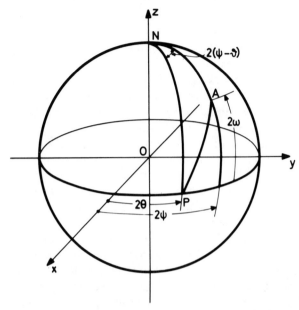

FIG. 9. Representation on the Poincaré sphere of the action of a linear polariser P on an elliptically polarised light beam A. The transmitted light intensity I through the polariser is equal to $I = \cos^2 (\widehat{AP}/2)$.

The eigenvector of the linear polariser will be presented on the Poincaré sphere by point P on the equator defined by the longitude 2θ (Fig. 9). The incident elliptically polarised light is represented by point A with longitude 2ψ and latitude 2ω. From the spherical triangle NPA of Fig. 9 it can be deduced that:

$$\cos \widehat{AP} = \cos 2\omega \cos 2(\psi - \theta)$$

so that relation (52) becomes:

$$I = \cos^2 (\widehat{AP}/2) \tag{53}$$

Relation (53) suggests that the light intensity of polarisation state A which is transmitted by a linear polariser whose eigenvector is defined by the polarisation state P is equal to square of the cosine of the half arc joining points A and P on the Poincaré sphere. Furthermore, it may be proved that this result is still valid for the more general case when the polariser is elliptical and it is represented on the Poincaré sphere by a point $P(2\theta, 2\varphi)$ not lying on its equatorial plane.

Another interesting result can be derived from eqn. (53) that the locus of the polarisation states on the Poincaré sphere, for which the emerging light intensity from a given polariser is constant, is a circle of the sphere with as pole the point representing the eigenvector of the polariser.

Moreover, the light intensity transmitted by a pair of crossed polarisers, that is polarisers whose eigenvectors represent orthogonal polarisation forms, is zero, since orthogonal forms are represented on the sphere by two opposite points A and P with $\widehat{AP} = \pi$.

Finally, it can be proved that a polariser of polarisation state A completely transmits the light of the same state A since $\widehat{AA} = 0$ and therefore $I = 1$. Moreover, a linear polariser P of an arbitrary azimuth transmits half the intensity of a circularly polarised beam (points N or S on the sphere) since for every point P on the equator of the sphere it is valid that $\widehat{PN} = \widehat{PS} = \pi/2$ and therefore $I = 1/2$.

We shall now establish the geometrical operation on the Poincaré sphere for defining the effect of a linear retarder of retardation δ whose fast axis subtends an angle θ with the Ox-axis of an elliptically polarised light beam. First we consider the simpler case when the incident light beam is linearly polarised at an angle i with the Ox-axis and the fast and slow axes of the retarder are along the Ox- and Oy-axes ($\theta = 0$) and afterwards we can extend the study to the more general case when the principal axes of the retarder do not coincide with the Ox- and Oy-axes and the retarder is elliptical.

From relation (14) we have for the angle i that:

$$\cos 2i = \cos 2\omega \cos 2\psi \tag{54}$$

Furthermore, from relations (11) and (13) it can be derived for the phase difference δ that:

$$\tan \delta = \frac{\tan 2\omega}{\sin 2\psi} \tag{55}$$

Let us now represent on the Poincaré sphere of Fig. 10 the incident on the linear retarder linearly polarised light by point P_0 such that $\widehat{FP_0} = 2i$ on the equator of the sphere and point F representing the horizontally linearly polarised light beam. The emerging light from the retarder light beam is represented by point $P(2\psi, 2\omega)$. From the right-angle spherical triangles FAP and FP_0P it may be derived by using relations (54) and (55) that:

$$\widehat{FP} = \widehat{FP_0} \quad \text{and} \quad \widehat{PFP_0} = \delta$$

This indicates that point P of the emerging light is obtained from P_0 of the

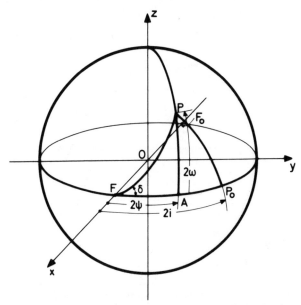

FIG. 10. Representation on the Poincaré sphere of the effect of a linear retarder with a retardation δ and whose eigenvectors F, S are along the positive and negative Ox axis on a linearly polarised light beam P_0. Point P of the emerging polarisation form is obtained from point P_0 of the incident form, by rotating the sphere around the FS axis by an angle $\widehat{P_0P} = \delta$.

incident light by rotating the Poincaré sphere about the FF_0-axis (Ox-axis) representing the polarisation form of the fast and slow eigenvectors of the linear retarder. The angle of rotation of the sphere is equal to angle δ of retardation of the retarder and the sense of rotation is counterclockwise for an observer looking from F at the centre of the sphere.

Since the definition of the horizontally linear polarised light beam on the Poincaré sphere determined by the position of Ox-axis, is arbitrary, the above result can be easily generalised to the case when the principal axes of the retarder are not along the Ox- and Oy-axes. Similarly, the above conclusions are still valid for the case when the incident light beam is not linearly polarised and we have the more general case of an elliptical retarder. Thus, in order to define the effect of an elliptical retarder on an elliptically polarised light beam the following operations should be made:

(i)   We define point P_0 on the sphere representing the incident light beam and points F and S representing the eigenvectors of the fast and slow axes of the retarder (Fig. 11). These points are

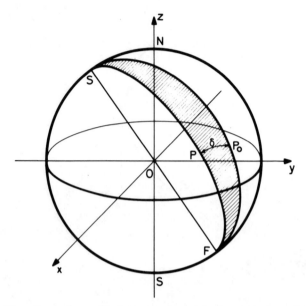

FIG. 11. The effect of an elliptical retarder with fast and slow eigenvectors represented by the points F and S on the Poincaré sphere and with retardation δ on an elliptically polarised light P_0. This effect is obtained by rotating the sphere around the axis FS at an angle δ and in a counterclockwise direction for an observer looking from F to S.

       diametrically opposite, since they represent a pair of orthogonal polarisation forms.

(ii)   We rotate the sphere about the FS-axis through an angle equal to the retardation δ in a sense counterclockwise for an observer looking from F to S.

(iii)   We find the new position P representing the light beam emerging from the retarder.

(iv)   In the case of a linear retarder, diameter FS lies on the equator of the Poincaré sphere.

(v)   A circular retarder has eigenvectors represented by the two poles (N and S) and therefore it converts any elliptical polarisation form to another one with the same handedness and the same ellipticity, and therefore every linear polarisation form is transformed to another linear polarisation form.

(vi)   A circularly polarised light represented by either of the poles N or S of the sphere is converted by means of a quarter-wave plate

($\delta = 90°$) into a linearly polarised light. The inverse is not always true, that is a linearly polarised light is not always converted to a circular one by a quarter-wave plate. To achieve this it is necessary that the axes of the plate and the polarisation form of the incident light are represented on the sphere by points which are in directions normal to each other, which means that the direction of polarisation must subtend an angle of 45° with the eigenvectors of the quarter-wave plate.

(vii) Insertion into any type of polarisation form of a 180° retarder (generally elliptical) has as a result to bring the polarisation form to its symmetric relative to the axis of the retarder.

(viii) For a train of retarders interposed in a polarised light beam the emerging light from the last retarder can be found from the incident light beam by defining the eigenvectors of the retarders on the Poincaré sphere and performing successive rotations of the sphere about axes defined by the particular eigenvectors of each retarder. The whole series of these retarders is equivalent to a single retarder whose retardation is equal to the arc of a great circle in the Poincaré sphere passing through the initial and final points of the successive retarders.

Interrelation Between the Mueller and Jones Matrices

Since the Mueller and Jones matrices represent optical devices and the corresponding Stokes and Jones vectors of the incident and emerging light beams are closely related, it follows that a relationship between the Mueller and Jones matrices must exist. We shall present without proof the relations existing between the elements of the two types of matrices.

Let us denote by S and S' the incident and emerging Stokes vectors of a certain polarisation form and a and a' the corresponding Jones vectors, and M and J the corresponding Mueller and Jones matrices of the optical element interposed in the polarised light. Then, the following relations hold:

$$\mathbf{S}' = \begin{bmatrix} s'_0 \\ s'_1 \\ s'_2 \\ s'_3 \end{bmatrix} = \begin{bmatrix} m_{11} & m_{12} & m_{13} & m_{14} \\ m_{21} & m_{22} & m_{23} & m_{24} \\ m_{31} & m_{32} & m_{33} & m_{34} \\ m_{41} & m_{42} & m_{43} & m_{44} \end{bmatrix} \begin{bmatrix} s_0 \\ s_1 \\ s_2 \\ s_3 \end{bmatrix} = \mathbf{MS} \qquad (56)$$

and:

$$\mathbf{a}' = \begin{bmatrix} A'_x \exp{(i\delta'_x)} \\ A'_y \exp{(i\delta'_y)} \end{bmatrix} = \begin{bmatrix} j_{11} & j_{12} \\ j_{21} & j_{22} \end{bmatrix} \begin{bmatrix} A_x \exp{(i\delta_x)} \\ A_y \exp{(i\delta_y)} \end{bmatrix} = \mathbf{Ja} \qquad (57)$$

It can be shown after some algebra and using previously established relations that the following formulas are valid between the coefficients m_{ij} $(i, j = 1, 2, 3, 4)$ and j_{kl} $(k, l = 1, 2)$ of the two matrices:

$$2|j_{11}|^2 = m_{11} + m_{12} + m_{21} + m_{22}$$

$$2|j_{21}|^2 = m_{11} + m_{12} - m_{21} - m_{22}$$

$$2|j_{12}|^2 = m_{11} - m_{12} + m_{21} - m_{22}$$

$$2|j_{22}|^2 = m_{11} - m_{12} - m_{21} + m_{22} \tag{58}$$

Relations (58) express the absolute values j_{kl} of the complex elements j_{kl} of the Jones matrix in terms of the elements m_{ij} of the corresponding Mueller matrix.

Furthermore, the polar angles θ_{kl} of the complex elements j_{kl} are similarly expressed by the relations:

$$\cos(\theta_{11} = \theta_{12}) = \frac{(m_{13} + m_{23})}{[(m_{11} + m_{21})^2 - (m_{12} + m_{22})^2]^{1/2}}$$

$$\sin(\theta_{11} - \theta_{12}) = \frac{(m_{14} + m_{24})}{[(m_{11} + m_{21})^2 - (m_{12} + m_{22})^2]^{1/2}}$$

$$\cos(\theta_{21} - \theta_{11}) = \frac{(m_{31} + m_{32})}{[(m_{11} + m_{12})^2 - (m_{21} + m_{22})^2]^{1/2}}$$

$$\sin(\theta_{21} - \theta_{11}) = \frac{(m_{41} + m_{42})}{[(m_{11} + m_{12})^2 - (m_{21} + m_{12})^2]^{1/2}}$$

$$\cos(\theta_{22} - \theta_{11}) = \frac{(m_{32} + m_{44})}{[(m_{11} + m_{22})^2 - (m_{21} + m_{12})^2]^{1/2}}$$

$$\sin(\theta_{22} - \theta_{11}) = \frac{(m_{43} + m_{34})}{[(m_{11} + m_{22})^2 - (m_{21} + m_{12})^2]^{1/2}} \tag{59}$$

Relations (58) and (59) completely determine the elements j_{kl} of the Jones matrices in terms of the elements m_{kl} of the corresponding Mueller matrices and the angles $(\theta_{11} - \theta_{12})$, $(\theta_{21} - \theta_{11})$ and $(\theta_{22} - \theta_{11})$ in the interval $0 < (\theta_{ij} - \theta_{kl}) < 2\pi$.

THE EQUIVALENCE THEOREM IN POLARISATION OPTICS

We shall now prove some theorems related to the behaviour of a train of retarders based on Jones calculus. Let us consider a polarised light beam with Jones vector $\mathbf{a}_0$ passing through a train of n optical retarders whose Jones matrices are $\mathbf{J}_n$ respectively $(n = 1, 2, 3 \ldots n)$. The Jones vector $\mathbf{a}$ of the light beam emerging from the above elements will be given by:

$$\mathbf{a} = \mathbf{J}_n \mathbf{J}_{n-1} \ldots \mathbf{J}_2 \mathbf{J}_1 \mathbf{a}_0$$

It has been already shown in relation (41) that the Jones matrix $\mathbf{J}(\theta)$ of a retarder is expressed by:

$$\mathbf{J}(\theta) = \mathbf{R}(-\theta)\mathbf{J}(0)\,\mathbf{R}(\theta) \tag{60}$$

where θ denotes the orientation of the fast axis of the retarder with respect to Ox-axis and $\mathbf{R}(\theta)$ is the rotation matrix.

If the orientation of the fast axes of retarders $1, 2, \ldots, (n-1), n$, with Jones matrices $\mathbf{J}_1, \mathbf{J}_2, \ldots, \mathbf{J}_{(n-1)}$, $\mathbf{J}_n$, with respect to Ox-axis are $\theta_1, \theta_2, \ldots, \theta_{(n-1)}, \theta_n$ respectively, the Jones vector $\mathbf{a}$ of the emerging light is expressed by:

$$\mathbf{a} = [\mathbf{R}(-\theta_n)\mathbf{J}_n(0)\mathbf{R}(\theta_n)][\mathbf{R}(-\theta_{(n-1)})\mathbf{J}_{n-1}(0)\mathbf{R}(\theta_{(n-1)})] \ldots$$
$$[\mathbf{R}(-\theta_2)\mathbf{J}_2(0)\mathbf{R}(\theta_2)][\mathbf{R}(-\theta_1)\mathbf{J}_1(0)\mathbf{R}(\theta_1)]\mathbf{a}_0$$

and by using the associative property of matrices we obtain for $\mathbf{a}$ that:

$$\mathbf{a} = \mathbf{R}(-\theta_n)\mathbf{J}\mathbf{R}(\theta_n)\mathbf{a}_0 = \mathbf{R}(-\theta_1)\mathbf{J}'\mathbf{R}(\theta_1)\mathbf{a}_0 \tag{61}$$

where matrices $\mathbf{J}$ and $\mathbf{J}'$ depend only on the relative orientation of the elements of the optical system and are independent of the orientation of the optical system as a whole and they are given by the relations:

$$\mathbf{J} = [\mathbf{J}_n(0)\mathbf{R}(\theta_n - \theta_{n-1})\mathbf{J}_{n-1}(0)\mathbf{R}(\theta_{n-1} - \theta_{n-2}) \ldots$$
$$\mathbf{J}_2(0)\mathbf{R}(\theta_2 - \theta_1)\mathbf{J}_1(0)\mathbf{R}(\theta_1 - \theta_n)] \tag{62a}$$

and

$$\mathbf{J}' = [\mathbf{R}(\theta_1 - \theta_n)\mathbf{J}_n(0)\mathbf{R}(\theta_n - \theta_{n-1})\mathbf{J}_{n-1}(0)\mathbf{R}(\theta_{n-1} - \theta_{n-2}) \ldots$$
$$\mathbf{J}_2(0)\mathbf{R}(\theta_2 - \theta_1)\mathbf{J}_1(0)] \tag{62b}$$

Relations (62) indicate the following theorem:

'The matrix representing a train of retarders is expressed in terms of two matrices the first of which depends only on the nature of the retarders and on their relative orientation, while the second matrix depends only on the orientation of the optical system as a whole with respect to the Ox-axis.'

This theorem may be considered as a generalisation of relation (60) valid for a single retarder for the case of a train of retarders.

It can be readily shown from relation (43) that the matrix of a retarder is unitary since it is the product of the unitary matrices $\mathbf{J}(0)$ and $\mathbf{R}(\theta)$. Since the product of unitary matrices is also a unitary matrix, the following theorem can be derived from relations (61) and (62):

'The Jones matrix representing a train of retarders is a unitary matrix'.

By applying the above theorem and the theorem that a unitary matrix may be considered as the product of a unitary matrix multiplied by a rotation matrix the following basic theorem may be derived:

'An optical system containing any number of retarders is optically equivalent to a system containing one retarder and one rotator'

This theorem constitutes the so-called *equivalence theorem* in polarisation optics.

From the nature of an optical rotator (crystal or solution presenting optical activity) it is deduced that the rotation of the polarisation plane introduced by the rotator changes sign when the light passes through the rotator in the reverse direction. Thus, if:

$$\mathbf{R}(\theta) = \begin{bmatrix} \cos\theta & \sin\theta \\ -\sin\theta & \cos\theta \end{bmatrix}$$

is the rotation matrix for the light passing through a rotator in a certain direction then the corresponding matrix for the light passing the same rotator but in the reverse direction is given by:

$$\mathbf{R}(-\theta) = \begin{bmatrix} \cos\theta & -\sin\theta \\ \sin\theta & \cos\theta \end{bmatrix}$$

These two last relations indicate that, since $\mathbf{R}(-\theta) = \mathbf{R}'(\theta)$ the rotation matrix corresponding to the light passing in the reverse direction is expressed by the transpose of the rotation matrix corresponding to the light passing in the normal direction. But, it has been established previously that for a polariser or a retarder the Jones matrix does not change if the light passes through them in the reverse direction. From relations (40) and (43) it can be also derived that the transpose matrices of the matrices representing either a polariser or a retarder are equal to the same matrices. Furthermore, according to the equivalence theorem, the Jones matrix $\mathbf{J}(\theta)$ of a given system including retarders can be expressed by:

$$\mathbf{J}(\theta) = \mathbf{J}_i(\theta)\mathbf{R}_i(\theta) \tag{63}$$

where $\mathbf{J}_i(\theta)$ and $\mathbf{R}_i(\theta)$ are the Jones matrices of the equivalent retarder and rotator respectively.

If the same optical system is traversed by the same beam but in the opposite direction, then its Jones matrix $\mathbf{J}^*(\theta)$, according to the previously established fundamental properties of the Jones matrices for a polariser and a retarder as well as for a rotator, will be given by:

$$\mathbf{J}^*(\theta) = \mathbf{R}_i(-\theta)\mathbf{J}_i(\theta) \qquad (64)$$

Relation (64) by taking into account that $\mathbf{R}'(\theta) = \mathbf{R}(-\theta)$ and further that $\mathbf{J}'_i(\theta) = \mathbf{J}_i(\theta)$ may be written as:

$$\mathbf{J}^* = \mathbf{R}'_i(\theta)\mathbf{J}'_i(\theta) \qquad (65)$$

Then, by taking the transpose of both sides of relation (63) we have that:

$$\mathbf{J}'(\theta) = \mathbf{R}'_i(\theta)\mathbf{J}'_i(\theta) \qquad (66)$$

From relations (65) and (66) it may be readily concluded that:

$$\mathbf{J}^*(\theta) = \mathbf{J}'(\theta) \qquad (67)$$

Relation (67) expresses the following theorem:

'The Jones matrix of an optical system traversed by a light beam in a certain direction must be transposed in order to obtain the Jones matrix of the same optical system traversed by the same beam but in the reverse direction'.

This theorem constitutes the *reversibility theorem* of polarisation optics.

MEASUREMENT OF ELLIPTICALLY POLARISED LIGHT

We shall here describe concrete methods for measuring the state of polarisation important for the photoelastic determination of stresses induced in a two- or three-dimensional body. In these problems the state of stress of the body modifies the polarisation form of the incident light beam so that the emerging light contains enough information for the determination of the polarisation state of the optical element which constitutes the model of the body.

For the complete characterisation of the polarisation state of an arbitrary light beam it is necessary to determine four independent quantities and as such we consider the four Stokes parameters. In the case of a perfectly polarised light beam three independent quantities characterise completely the polarisation state. Indeed, for the case when the Stokes parameters are used for defining a polarised light beam only three independent quantities are necessary since the four parameters must always satisfy the initial condition expressed in relation (16). These three independent Stokes parameters can be used to specify the polarisation form also in terms of the Jones vector. Generally, these three quantities may be selected to represent the two amplitudes and the phase difference of the components of the light vector along any two orthogonal directions.

When the light ellipse is used to define the polarisation state and we are not interested in the absolute values of the intensities, the azimuth, ψ, the ellipticity, ω, and the handedness of rotation of the light vector may be considered the most convenient parameters to specify the polarisation state. In the graphical method of Poincaré sphere the two spherical coordinates of the points of the sphere and the selection of the point to lie either on the northern or on the southern hemisphere constitute the three convenient parameters for defining the polarisation state.

Since the light intensity is the most convenient quantity to be determined experimentally, all methods defining polarisation states are based on the optical transformation of the polarised light by inserting optical elements into the polarised light beam convenient for measuring the intensity of the emerging light. Such optical elements are usually used, either polarisers or retarders or both these devices. These elements can be combined in various systems depending on the particular method used for measuring the emerging polarisation form.

In the following we shall present the principles of the basic methods used for measuring a polarisation state. All methods which will be developed here are based on the same principle of insertion of polarisers and retarders into the light beam and the measurement of the intensity of the emerging light. Since the light emerging from a retarder is generally elliptically polarised, the final optical element of any measuring device should always be a linear polariser. Thus, the emerging light always has a definite direction for its light vector, easily detected by an intensity measuring device. In this case the polariser acquires the name of *analyser*. However, the insertion of a linear analyser does not suffice for the measurement of an elliptical polarisation form. Only linearly polarised light can be detected by an analyser alone. In all other cases a retarder is necessary to be interposed in the light path.

We try now to make a quantitative analysis of an elliptical polarisation form based on Mueller calculus. For this purpose we must insert a retarder and an analyser into the light path. Suppose that the Stokes vector **S** of the light beam to be measured is given by:

$$\mathbf{S} = \begin{bmatrix} s_0 \\ s_1 \\ s_2 \\ s_3 \end{bmatrix}$$

and let us insert a retarder of retardance δ with its eigenvector subtending an angle β with the Ox-axis and a linear analyser whose pass axis subtends an angle γ with the Ox-axis. Then, the Mueller matrices $\mathbf{R}_\beta(\delta)$ of the retarder and $\mathbf{P}_\gamma$ of the polariser are given by:

$$\mathbf{R}_\beta(\delta) = \begin{bmatrix} 1 & 0 & 0 & 0 \\ 0 & \cos^2 2\beta + \sin^2 2\beta \cos\delta & (1 - \cos\delta)\sin 2\beta \cos 2\beta & -\sin 2\beta \sin\delta \\ 0 & (1 - \cos\delta)\sin 2\beta \cos 2\beta & \sin^2 2\beta + \cos^2 2\beta \cos\delta & \cos 2\beta \sin\delta \\ 0 & \sin 2\beta \sin\delta & -\cos 2\beta \sin\delta & \cos\delta \end{bmatrix} \tag{68}$$

and:

$$\mathbf{P}_\gamma = 1/2 \begin{bmatrix} 1 & \cos 2\gamma & \sin 2\gamma & 0 \\ \cos 2\gamma & \cos^2 2\gamma & \sin 2\gamma \cos 2\gamma & 0 \\ \sin 2\gamma & \sin 2\gamma \cos 2\gamma & \sin^2 2\gamma & 0 \\ 0 & 0 & 0 & 0 \end{bmatrix} \tag{69}$$

The Stokes vector **S**′ of the emerging light from the analyser is given by:

$$\mathbf{S}' = \mathbf{P}_\gamma \mathbf{R}_\beta(\delta)\mathbf{S} \tag{70}$$

Relation (70) yields for the first element s_0' of the emerging light beam, which equals the emerging light intensity $I(\beta, \gamma, \delta)$, the relation:

$$2I(\beta, \gamma, \delta) = 2s_0' = s_0 + [s_1 \cos 2\beta + s_2 \sin 2\beta]\cos 2(\beta - \gamma)$$
$$+ [(s_1 \sin 2\beta - s_2 \cos 2\beta)\cos\delta - s_3 \sin\delta]\sin 2(\beta - \gamma) \tag{71}$$

Since the intensity depends on the three parameters s_1, s_2, s_3 (since $s_0^2 = s_1^2 + s_2^2 + s_3^2$) and on the orientation β and γ of the retarder and the analyser, as well as on the retardance δ, it can be evaluated in different ways by assigning three different sets of values to the parameters β, γ and δ and measuring the corresponding values of the emerging light intensity. Thus, the methods for determining the polarisation state may be classified into three categories; (i) those using a rotating analyser (γ variable β, δ constants), (ii) those using a rotating retarder (β variable γ, δ constants) and, (iii) those using a retarder of varying retardance δ, or using a series of retarders (δ variable, β, γ constants).

As an application of the above outlined principles let us determine the elements of Stokes and Jones vectors for a polarised light beam.

For the Stokes vector we can use relation (71) to determine the emerging light beam. For this purpose we interpose successively the following optical elements and we calculate the emerging light beams for each element:

(i)   A linear analyser with its pass axis parallel to Ox-axis. Then we have from eqn. (71) that:

$$I_1 = I(0,0,0) = \tfrac{1}{2}(s_0 + s_1) \tag{72}$$

(ii)   A linear analyser with its pass axis parallel to Oy-axis. Then:

$$I_2 = I(0,90°,0) = \tfrac{1}{2}(s_0 - s_1)$$

From the last two relations we easily obtain that:

$$s_0 = I_1 + I_2 \quad \text{and} \quad s_1 = I_1 - I_2 \tag{73}$$

(iii)   A linear analyser with its pass axis at 45° with the Ox-axis. Then:

$$I_3 = I(0,45°,0) = \tfrac{1}{2}(s_0 + s_2)$$

and:

$$s_2 = 2I_3 - (I_1 + I_2) \tag{74}$$

(iv)   A retarder with retardance $\delta = 90°$ (quarter-wave plate) and with its fast axis along the Ox-axis followed by an analyser with its pass axis at 45° to the Ox-axis ($\gamma = 45°$). Then it is valid:

$$I_4 = I(0,45°,90°) = \tfrac{1}{2}(s_0 + s_3)$$

and therefore:

$$s_3 = 2I_4 - (I_1 + I_2) \tag{75}$$

Thus, the four Stokes parameters may be evaluated from relations (72) to (75) by measuring the four intensities I_1, I_2, I_3 and I_4. Usually, in all measuring devices only the relative values of light intensities normalised to the intensity I_1 are measured and this makes easier and more accurate the experimental determination of the Stokes parameters since in this way any calibration of the measuring device can be avoided.

By using the same procedure and the same optical elements similarly arranged we can evaluate the components of the Jones vector. Thus:

(i)   For a linear analyser with its pass axis parallel to Ox-axis we have:

$$\mathbf{a'} = \mathbf{Ja} = \begin{bmatrix} 1 & 0 \\ 0 & 0 \end{bmatrix} \begin{bmatrix} a_x \\ a_y \end{bmatrix} = \begin{bmatrix} a_x \\ 0 \end{bmatrix}$$

where:

$$\mathbf{a} = \begin{bmatrix} a_x \\ a_y \end{bmatrix} = \begin{bmatrix} A_x \exp{(i\delta_x)} \\ A_y \exp{(i\delta_y)} \end{bmatrix} = \exp{(i\delta_x)} \begin{bmatrix} A_x \\ A_y \exp{(i\delta)} \end{bmatrix}, \delta = (\delta_y - \delta_x)$$

is the Jones vector of the impinging light. The intensity I_1 of the transmitted light is given by:

$$I_1 = [a_x \quad 0] \begin{bmatrix} a_x \\ 0 \end{bmatrix} = A_x^2 \tag{76}$$

(ii)   for a linear analyser with its pass axis parallel to Oy-axis we have:

$$\mathbf{a'} = \mathbf{Ja} = \begin{bmatrix} 0 & 0 \\ 0 & 1 \end{bmatrix} \begin{bmatrix} a_x \\ a_y \end{bmatrix} = \begin{bmatrix} 0 \\ a_y \end{bmatrix}$$

and:

$$I_2 = [0 \quad A_y \exp{(-i\delta)}] \begin{bmatrix} 0 \\ A_y \exp{(i\delta)} \end{bmatrix} = A_y^2 \tag{77}$$

(iii) For a linear analyser with its pass axis at 45° to Ox-axis we have:

$$\mathbf{a'} = \mathbf{Ja} = \frac{1}{2} \begin{bmatrix} 1 & 1 \\ 1 & 1 \end{bmatrix} \begin{bmatrix} a_x \\ a_y \end{bmatrix} = \frac{1}{2} \begin{bmatrix} a_x + a_y \\ a_x + a_y \end{bmatrix}$$

and the intensity I_3 is given by:

$$I_3 = \tfrac{1}{4}[A_x + A_y \exp{(-i\delta)} A_x + A_y \exp{(-i\delta)}] \begin{bmatrix} A_x + A_y \exp{(i\delta)} \\ A_x + A_y \exp{(i\delta)} \end{bmatrix}$$

or:

$$I_3 = \tfrac{1}{2}[A_x^2 + A_y^2 + 2A_x A_y \cos{\delta}] \tag{78}$$

(iv)   for a retarder ($\delta = 90°$) with its fast axis along the Ox- axis followed by a linear analyser with its pass axis at 45° to Ox-axis we have:

$$\mathbf{a'} = \mathbf{J_2 J_1 a} = \frac{1}{2} \begin{bmatrix} 1 & 1 \\ 1 & 1 \end{bmatrix} \begin{bmatrix} i & 0 \\ 0 & 1 \end{bmatrix} \begin{bmatrix} a_x \\ a_y \end{bmatrix} = \frac{1}{2} \begin{bmatrix} ia_x + a_y \\ ia_x + a_y \end{bmatrix}$$

and:

$$I_4 = \tfrac{1}{4}[-iA_x + A_y \exp(-i\delta) - iA_x + A_y \exp(-i\delta)]$$
$$\times \begin{bmatrix} iA_x + A_y \exp(i\delta) \\ iA_x + A_y \exp(i\delta) \end{bmatrix}$$

or:

$$I_4 = \tfrac{1}{2}(A_x^2 + A_y^2 + 2A_x A_y \sin \delta) \tag{79}$$

By measuring the light intensities I_1 and I_2 we evaluate the amplitudes A_x and A_y, while from the intensities I_3 and I_4 we completely determine the retardance δ.

Photoelectric Methods for Measuring the State of Polarisation

It was already mentioned that the determination of the polarisation state of a light beam is greatly simplified by interposing a retarder followed by an analyser.

The light intensity relation given in eqn. (71) for the first and third methods where either a rotating analyser or a retarder of variable retardance is used can be put in the general form:

$$I(\varphi) = c_1 + c_2 \cos(\varphi + c_3) \tag{80a}$$

while the second method where a rotating retarder is used the general expression for the intensity takes the form:

$$I(\varphi) = c_1 + c_2 \cos(\varphi + c_3) + c_4 \cos 2(\varphi + c_5) \tag{80b}$$

In relations (80) φ being the variable parameter of the system may be identified each time to one of the quantities γ, β and δ for the first, second and third method respectively, while c_i ($i = 1, 2, 3, 4, 5$) are constants depending on Stokes parameters and on those values of γ, β and δ which for each particular experiment have been kept constant.

As an example we consider the case where a retarder with $\delta = 90°$ (quarter-wave plate) and an analyser whose pass axis lies along the Ox-axis ($\gamma = 0$) are interposed in the light beam. Then relation (71) becomes:

$$2I(\beta) = (S_0 + S_1/2) - s_3 \sin 2\beta + \tfrac{1}{2}(s_1^2 + s_2^2)^{1/2} \cos(4\beta - \tan^{-1} s_2/s_1) \tag{81}$$

For a perfectly polarised light beam and for the case of absolute measurements of the intensity of the emerging light either relation (80a) or relation (80b) suffices for the determination of the three unknown quantities of the light beam.

For the case when it is not known that the light beam is perfectly polarised or for the case when the relative values of the emerging light intensities are sought the first or the third of methods leading to eqn. (80a), do not suffice alone to determine the polarisation state. In this case the second method using a rotating retarder and leading to eqn. (80b) is convenient.

By using any type of detecting instrument of light intensity the output signal yields the value of light intensity provided the calibration constant of the instrument is known. However, for relative values of light intensity there is no need for a calibration test. The signal of the detector can be further treated to yield directly the constants c_i of relations (80).

The techniques used to determine the constants c_i of relations (80) can be classified into the following categories: (i) in the first category belong all the techniques where two intensity measurements are performed and the value of phase ($\varphi + c_3$) corresponding to a particular point of the output signal is determined. (ii) The second category includes all methods where the light intensity at three different values of angle φ is measured. (iii) The third category comprises all the methods where the respective parameter of the system linearly varying with time, as well as the constant term c_1, and the amplitudes c_2, c_4 and the corresponding phases c_3 and c_5 of the oscillatory components of the emergent signal are separately determined.

This method is self-sufficient for the general case of an elliptically polarised light.

We shall describe now some special techniques for determining the polarisation state of an emerging light beam. The first technique which is due to Kent and Lawson[39] is a photoelectric technique based on the fact that, when a circularly polarised light passes through a rotating analyser the output signal does not present any oscillatory component. Thus, the method consists in converting a given elliptical polarisation form into a circularly polarised form through a suitably selected and oriented retarder and afterwards in passing the outcoming light through a rotating analyser. Thus, the incident to the retarder polarised light is presented by point $A(2\psi, 2\omega)$ on the Poincaré sphere (Fig. 12).

In order to bring point A to point N representing the right circularly polarised light on the sphere, the Poincaré sphere must be rotated about the BC-axis lying on the equatorial plane with $B(270° + 2\psi, 0)$ and $C(90 + 2\psi, 0)$, through an angle φ such that

$$|2\omega| + \varphi = 90°$$

Thus we obtain for the angular coordinates of the original beam that

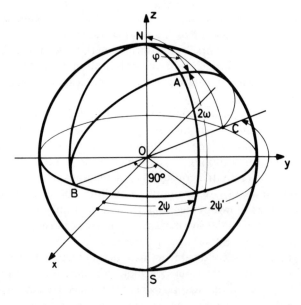

FIG. 12.   Determination of the orientation of the principal axes and the retardance
of a retardation plate which converts a given elliptical polarisation form into a
circular one by the Poincaré sphere method. Point A represents the given
polarisation form, points B and C the eigenvectors and φ the retardance of the
retardation plate.

$2\psi = (2\psi' - 90°)$ and $|2\omega| = (90° - \varphi)$ where $2\psi'$ is the longitude of point
C.

However, the handedness of the light ellipse cannot be determined since
there is no way to distinguish whether the induced circularly polarised light
is right or left.

A second double-beam technique was developed by Archard and co-
workers.[40] In this method a double image polariser was used to separate the
two mutually perpendicular linear vibrations into two 90° out-of-phase
components. When these two components become of equal intensity the
eigenvectors of the rotating polariser are at 45° to the eigenvectors of the
light ellipse and hence the azimuth of the ellipse may be determined.
Subsequently the polariser is rotated by an angle of 45° so that its axes lie
now along the principal axes of the light ellipse. The ratio of intensities of
the two signals allows the evaluation of the ellipticity of the light ellipse.

A third technique introduced by Robert[41-43] utilises a rotating analyser
which detects the characteristics of the output signal. In this case eqn. (71)

under the same conditions as previously takes the form:

$$I(0, \gamma, 0) = \tfrac{1}{2}[s_0 + (s_1^2 + s_2^2)^{1/2} \cos(2\gamma - \tan^{-1} s_2/s_1)]$$

and consists of a constant term and an oscillatory one with frequency equal to twice the frequency of the rotating analyser. By measuring the relative values of the amplitudes of the constant and oscillatory terms, as well as the phase of the oscillatory term, two out of the three Stokes parameters can be evaluated from which the azimuth and the ellipticity of the light ellipse can be immediately determined. In order to determine the handedness of the ellipse a 90° retarder with its fast axis parallel to the major axis of the ellipse is placed before the analyser. The phase of the oscillatory component gives the required handedness. Robert has constructed an automatic photoelasticimeter based on this method.

Another possibility introduced by Sekera[44] is to use a polarimeter for detecting the three amplitudes and the two phase differences of relation (81) for the complete characterisation of a polarisation form.

Visual Methods for Measuring the Polarisation State

This group of methods includes techniques where optical elements are inserted in the light path and by a suitable arrangement either the emerging light intensity becomes an extremum, or two light intensities emerging from the optical device turn out to be equal. We shall describe techniques for determining the azimuth, the ellipticity and the handedness of the light ellipse.

In order to determine the azimuth of a light ellipse we utilise a rotating analyser. When the pass axis of the analyser is parallel to the major or minor axes of the light ellipse the intensity of light transmitted becomes maximum or minimum. This phenomenon may be immediately induced in the Poincaré sphere. If the polarisation state is represented by P (Fig. 13) and the linear analyser by A then it is valid that:

$$I = \cos^2 \frac{\widehat{PA}}{2}$$

The intensity I passes through an extremum when point A lies on the same meridian as point P, namely at positions A_{max} and A_{min}. Since the minimum value for I is more easily detected than the maximum we prefer to express the azimuth ψ_p of the light ellipse by:

$$\psi_p = (\psi_{A_{min}} - \pi/2)$$

However, in practice the exact position of the analyser for minimum

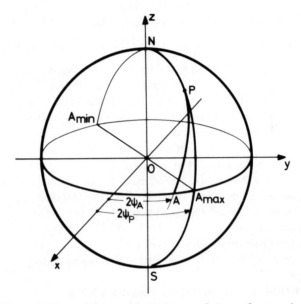

Fig. 13. Determination of the positions A_{max} and A_{min} of an analyser which transmits the maximum and minimum light intensity of an incident polarisation form represented by the point P on the Poincaré sphere.

transmitted intensity cannot be determined accurately. On the other hand, the accuracy is strongly reduced when the semi-axes of the ellipse tend to become equal. For these reasons all methods for determining the azimuth are based on the separation of the field of view into two parts and the comparison of their intensities.

Many special techniques using various optical devices have been conceived to materialise the principle of dividing the field of view into two parts. Among these we mention without any further description the method of double-field analyser[45-48] the method of the rotating bi-plate[49-51] and the method of half-shadow plate.[52-53] The details of the description of the principles and the optical devices used in these methods are described in the above-mentioned references (see also reference 58).

For the measurement of the ellipticity of polarised light, use is normally made of a retarder (either a half-, or a quarter-wave, or any retardance plate) interposed in the light beam under measurement. Two different approaches exist. In the first approach the retardance is constant and the ellipticity is measured by varying the orientation of the retarder while in the second approach the retarder lies in a fixed orientation but with a variable retardance. The methods belonging to the first approach are based on the Senarmont principle,[54] while those of the second approach use various types of compensators.

Generally speaking the methods for measuring the ellipticity presume the knowledge of the azimuth although in some methods both the azimuth and the ellipticity are simultaneously determined. The evaluation of ellipticity is of particular importance for the determination of the retardation of a birefringent medium which in succession gives the stress distribution in the model.

Besides the above classification of the methods in photoelectric and visual methods they can also be divided, by means of the approach used in measuring the polarisation state, into direct and indirect methods. In the direct methods the optical elements are suitably arranged to measure some characteristic points of the light ellipse, whereas in the indirect methods the light ellipse is suitably transformed by adjusting the optical elements to measure some known polarisation forms.

The direct methods are based on the Senarmont principle according to which, when a 90° retarder is placed with its principal axes parallel to those of the light ellipse of an elliptically polarised light, the emerging light is linearly polarised at azimuth ω with respect to the slow axis of the incident light of ellipticity ω. Figure 14 represents graphically the procedure for the Senarmont method. Point P on the Poincaré sphere represents the incident light, while points M and M' the fast and slow axes of the 90° retarder (quarter-wave plate). Point M' lies on the same meridional as point P and points M and M' on a diameter of the equator. The emerging light is obtained by rotating the sphere about the MM'-axis at an angle of 90°. This rotation brings P to P' with $\widehat{M'P} = \widehat{M'P'}$ that is the azimuth of the emerging polarised (plane-polarised) light with respect to the slow axis of the 90° retarder equals the ellipticity of the incident light.

The application of the Senarmont principle in measuring the ellipticity is strongly dependent on the correct orientation of the 90° retarder which must be parallel to the principal axes of the incident light ellipse. If this is not achieved it is impossible to obtain complete extinction of the emerging light from the analyser.

Various other methods were developed for overpassing this disadvantage of the Senarmont method. These are either successive approximation methods.[55,56] or methods based on the half-shade principle.[57]

The determination of both the azimuth and the ellipticity of a light beam can also be achieved by placing two crossed polarisers and two 90° retarders in the light path and detecting the emerging light. This arrangement

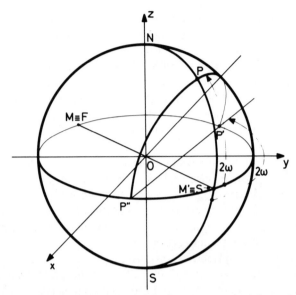

FIG. 14. Geometrical representation of the Senarmont principle on the Poincaré sphere. An elliptically polarised light P whose axes of the light ellipse are parallel to the axes M, M' of a quarter-wave plate is converted by the plate into a linearly polarised light P' at an azimuth ω to the slow axis of the plate equal to the ellipticity of the elliptically polarised light.

constitutes the basic set-up for the photoelastic method of stress analysis and is studied separately in the previous chapter.

In order to define the handedness of the polarisation form it is possible to use either the methods for measuring the ellipticity of the polarised light or by using compensation methods. When a 90° retarder is interposed in the polarised light beam the handedness of the light ellipse is right when the fast axis of the retarder is parallel to the major axis of the light ellipse and the analyser is rotated by an angle γ varying between $0 < \gamma < 180°$ until an extinction of emerging light is obtained. On the other hand, similar principles hold for the cases when various types of compensators are used but these methods are described in the previous chapter dealing with the methods of photoelasticity.

Determination of Matrix Elements of an Optical System

Any optical system composed either of single retarders and polarisers or of a pile of such elements can be completely characterised by its corresponding Mueller or Jones matrix. On the other hand, knowledge of the matrix of a system allows the evaluation of the Stokes or Jones vector of an emerging form from the optical system light, if the incident light is known.

The determination of the elements of the Mueller or Jones matrix of an optical system can be achieved by letting various typical light forms to pass through the system and determine the corresponding Stokes or Jones vectors of the emerging light. In the following we shall examine how we can determine experimentally the elements of a Mueller or Jones matrix.

Let us assume the Mueller matrix $\mathbf{M}$ of an optical element given by:

$$\mathbf{M} = \begin{bmatrix} m_{11} & m_{12} & m_{13} & m_{14} \\ m_{21} & m_{22} & m_{23} & m_{24} \\ m_{31} & m_{32} & m_{33} & m_{34} \\ m_{41} & m_{42} & m_{43} & m_{44} \end{bmatrix}$$

whose elements m_{ij} $(i, j = 1, 2, 3, 4)$ must be determined. For this purpose we assume that different light forms impinge on the system and we can completely determine the Stokes vector of the emerging light form in each case. As such cases they are considered the following:

(i) A unit-intensity unpolarised light with Stokes vector $\mathbf{S}_1$ given by:

$$\mathbf{S}_1 = \begin{bmatrix} 1 \\ 0 \\ 0 \\ 0 \end{bmatrix}$$

The emerging Stokes vector $\mathbf{S}_1'$ is given by $\mathbf{S}_1' = \mathbf{M}\mathbf{S}_1$ with:

$$\mathbf{S}_1' = \begin{bmatrix} m_{11} \\ m_{21} \\ m_{31} \\ m_{41} \end{bmatrix}$$

Thus the elements m_{i1} $(i = 1, 2, 3, 4)$ of $\mathbf{M}$ are determined.

(ii) A linearly polarised light of unit intensity whose light vector is parallel to the Ox-axis. Then the Stokes vectors $\mathbf{S}_2$ and $\mathbf{S}_2'$ for the incident and emerging light beam are expressed by:

$$\mathbf{S}_2 = \begin{bmatrix} 1 \\ 1 \\ 0 \\ 0 \end{bmatrix}, \qquad \mathbf{S}_2' = \begin{bmatrix} m_{11} + m_{12} \\ m_{21} + m_{22} \\ m_{31} + m_{32} \\ m_{41} + m_{42} \end{bmatrix}$$

(iii) A linearly polarised light of unit intensity whose light vector subtends an angle of $45°$ with the Ox-axis. Then, the Stokes vectors S_3 and S_3' take the form:

$$S_3 = \begin{bmatrix} 1 \\ 0 \\ 1 \\ 0 \end{bmatrix}, \qquad S_3' = \begin{bmatrix} m_{11} + m_{13} \\ m_{21} + m_{23} \\ m_{31} + m_{33} \\ m_{41} + m_{43} \end{bmatrix}$$

(iv) A right-handed circularly polarised light of unit intensity. Then, the Stokes vectors S_4 and S_4' of the incident and emerging light take the form:

$$S_4 = \begin{bmatrix} 1 \\ 0 \\ 0 \\ 1 \end{bmatrix}, \qquad S_4' = \begin{bmatrix} m_{11} + m_{14} \\ m_{21} + m_{24} \\ m_{31} + m_{34} \\ m_{41} + m_{44} \end{bmatrix}$$

The last four pairs of values for the emerging Stokes vectors are sufficient for the evaluation of the elements m_{ij}.

For the case of determining the elements j_{kl} of the Jones matrix $J(k, l = 1, 2)$ we can proceed in the same way. We assume first that two linearly polarised light forms impinge on the element whose Jones matrix is expressed by:

$$J = \begin{bmatrix} j_{11} & j_{12} \\ j_{21} & j_{22} \end{bmatrix}$$

One of the two linearly polarised forms has a light vector parallel to the Ox-axis while the other is parallel to the Oy-axis. Then, for the first case we have that:

$$a_1' = Ja_1$$

with:

$$a_1 = \begin{bmatrix} 1 \\ 0 \end{bmatrix} \quad \text{and therefore:} \quad a_1' = \begin{bmatrix} j_{11} \\ j_{12} \end{bmatrix}$$

For the second case we have respectively:

$$a_2 = \begin{bmatrix} 0 \\ 1 \end{bmatrix} \quad \text{and therefore:} \quad a_2' = \begin{bmatrix} j_{12} \\ j_{22} \end{bmatrix}$$

This completes the evaluation of J_{kl} of the Jones matrix J.

CONCLUSIONS

In conclusion we have given an overall view of the modern methods based on matrix calculus, that is the Mueller and Jones calculi from the point of view of analytic confrontation of the problem and the Poincaré sphere representation from the point of view of graphical methods for the solution of problems of polarisation optics. The efficiency compactness and simplicity of the methods are self-evident throughout this chapter. We did not tackle the problem of application of these methods to two- and three-dimensional photoelasticity and its natural extension in holography and holographic interferometry. This subject deserves a special chapter by itself.

Moreover we did not go into details of methods based on the above fundamental methods because of lack of space although we have recognised their special features and capabilities. Thus, neither the method of quaternions, nor the j-circle method have been conveniently developed. For the first method the reader is referred to the scattered bibliography already existing and contained in reference 58. For the j-circle method a convenient and comprehensive book is that by Kuske[31] where the interested reader will find all the detail but it is not presented with a matrician calculus.

ACKNOWLEDGEMENT

This research work was partly supported by funds of the National Technical University of Athens. The author is indebted to his assistant Mr C. Spyropoulos for his help in reading the manuscript and for suggestions.

REFERENCES

 1. STOKES, C. G., *Trans. Cambridge Phil. Soc.*, 1852, **9,** 399.
 2. MUELLER, H., *J. Opt. Soc. Am.*, 1948, **38,** 661.
 3. JONES, R. C., *J. Opt. Soc. Am.*, 1941, **31,** 500.
 4. JONES, R. C., *J. Opt. Soc. Am.*, 1942, **32,** 486.
 5. JONES, R. C., *J. Opt. Soc. Am.*, 1947, **37,** 107.
 6. JONES, R. C., *J. Opt. Soc. Am.*, 1947, **37,** 110.
 7. JONES, R. C., *J. Opt. Soc. Am.*, 1948, **38,** 671.
 8. JONES, R. C., *J. Opt. Soc. Am.*, 1956, **46,** 126.
 9. JONES, R. C., *J. Opt. Soc. Am.*, 1956, **46,** 528.
10. WHITNEY, C., *J. Opt. Soc. Am.*, 1971, **61,** 1207.

11. RICHARTZ, M. and HSU, H.´Y., *J. Opt. Soc. Am.*, 1949, **39**, 136.
12. CERNOSEK, J., *J. Opt. Soc. Am.*, 1971, **61**, 324.
13. CERNOSEK, J., *Exp. Mech.*, 1973, **13**, 83.
14. CERNOSEK, J., *Exp. Mech.*, 1973, **13**, 273.
15. CERNOSEK, J., *Exp. Mech.*, 1975, **15**, 354.
16. POINCARÉ, H., '*Théorie Mathématique de la Lumiére*' Vol. 2, Chap. 12, Gauthiers-Villars Paris, 1892.
17. BECQUEREL, J., *Communs. Phys. Lab. Univ. Leiden*, 1928, No 91C.
18. BECQUEREL, J., *Communs. Phys. Lab. Univ. Leiden*, 1930, No 221A.
19. SKINNER, C. A., *J. Opt. Soc. Am.*, 1925, **10**, 490.
20. CHAUMONT, L., *C.R. Acad. Sci.*, 1913, **150**, 1604.
21. BRUAT, G. and GRIVET, P., *J. Phys. Radium*, 1935, **6**, 12.
22. BJORNSTAHL, Y., *Physik. Zeitschr.*, 1939, **40**, 437.
23. JERRARD, H. G., *J. Opt. Soc. Am.*, 1954, **44**, 634.
24. KOESTER, C. J., *J. Opt. Soc. Am.*, 1959, **49**, 405.
25. RAMACHANDRAN, G. N. and RAMASESHAN, S., 'Crystal Optics' in *Encyclopedia of Physics* (S. Flügge ed.), Springer, Berlin, 1961, 25/1,1.
26. PANCHARATNAM, S., *Proc. Ind. Acad. Sci.*, 1956, A43, 247.
27. PANCHARATNAM, S., *Proc. Ind. Acad. Sci.*, 1956, A44, 398.
28. ABEN, H. K., *Integrated Photoelasticity* Valgus, Tallin, USSR, 1975.
29. ROBERT, A., *Polarimétrie et Photoélasticimétrie* Serv. Techn. Const. Armes Navales, Paris, 1972.
30. MENGES, H. J., *Z. Angew. Math. Mech.*, 1940, **20**, 210.
31. KUSKE, A., *Einführung in die Spannungsoptik* Wissenchaftliche Verlagsgesellschaft, Stuttgart, 1959.
32. KUSKE, A. and ROBERTSON, G., *Photoelastic Stress Analysis*, John Wiley and Sons, London, 1974.
33. WRIGHT, F. E., *J. Opt. Soc. Am.*, 1930, **20**, 529.
34. ROBERT, A. J., *Exp. Mech.*, 1967, **7**, 224.
35. ABEN, H. K., *Proc. Conf. on Exp. Methods of Investigating Stress and Strain in Structures*, Praha, **33**, 1965.
36. ABEN, H. K., *Proc. Fourth Int. Conf. Stress Anal.* Cambridge, 1970.
37. SCHWIEGER, H., *Exp. Mech.*, 1969, **9**, 67.
38. COKER, E. G. and FILON, L. N. G., *A Treatise on Photo-elasticity* 2nd edn., University Press, Cambridge, 1957.
39. KENT, C. V. and LAWSON, J., *J. Opt. Soc. Am.*, 1973, **27**, 117.
40. ARCHARD, J. F., CLEGG, P. L. and TAYLOR, A. M., *Proc. Phys. Soc. London*, 1952, **B65**, 758.
41. ROBERT, A., BOURDON, C. and LE GOER, J. R., *Fr. Méc.*, 1967, **24**, 93.
42. ROBERT, A. and FERRE, M., *Bull. ATMA*, 1969.
43. ROBERT, A., *Int. J. Solids Struct.*, 1970, **6**, 423.
44. SEKERA, Z., *Adv. Geophys.*, 1956, **3**, 43.
45. JELLETT, J. H., *Rep. Brit. Assoc.*, 1860, **30**, 13.
46. CORNU, M. A., *Bull. Soc. Chim.*, 1870, **14**, 140.
47. SCHÖNROCK, O., *Handbuch der Physik*, 1928, **19**, 750.
48. LIPPICH, F., *Wien. Ber.*, 1885, **91**, 1059.
49. NAKAMURA, S., *Centralbltt für Min.*, 1905, 267.
50. BERTRAND, E., *Bull. Soc. Mineral.*, 1875, **1**, 22.

51. STRONG, J., *Rev. Sci. Inst.*, 1935, **6**, 243.
52. CHAUVIN, M., *Ann. de Toulouse*, 1889, **3**, 30.
53. CHAUMONT, M., *Ann. de Phys.*, 1915, **4**, 175.
54. FÖPPL, L. and MÖNCH, E., *Praktische Spannungsoptik*, Springer-Verlag Heidelberg, 1959.
55. STOKES, C. G., *Math. Phys.*, 1901, **3**, 197.
56. MACCULLAGH, J., *Collected Works*, London, 1880, **138**, 230.
57. RICHARTZ, M. Z., *Instrumentenk*, 1940, **60**, 357.
58. THEOCARIS, P. S. and GDOUTOS, E. E., *Matrix Theory of Photoelasticity*, Springer-Verlag, Heidelberg, 1979.

Chapter 4

STRAIN GAUGES

A. L. WINDOW

Welwyn Strain Measurement Ltd, Basingstoke, UK

SUMMARY

This chapter is a broad survey of the state of the art in strain gauge techniques, reviewing very briefly the available strain gauges in the wider meaning of the term, but concentrating heavily on the electrical resistance type which is the most widely used. The discussion of resistance strain gauges includes the resistance alloys and backing materials in common use, types of construction, and installation methods. Installation methods are sub-divided to cover surface preparation, adhesives, leadwire selection and attachment, and gauge protection. A brief review of types of instrumentation currently used with resistance strain gauges is followed by a review of performance limits and a reference to the analysis of experimental data from strain gauges.

INTRODUCTION

Strain gauges have been in use in experimental stress analysis for over forty years and are probably more widely used, understood, and accepted, than any other experimental technique. That is not to say that they are used widely enough, or understood well enough by all who use them. If this chapter, by making engineers aware of what can be done, encourages the wider use of experimental stress analysis in industry it will have served its purpose.

It is fair to say that in the last few years there have been no revolutionary inventions. What has occurred is a very important steady improvement in

127

materials and in techniques of both manufacture and installation which has made strain gauges far more reliable, accurate, and versatile. They should now be the everyday tool of all engineers concerned with design, and with prototype testing, as well as for failure analysis!

The design and performance of precision transducers such as load cells, and electronic scales is a subject beyond the scope of this chapter, nevertheless it is worth commenting that these often use the same strain gauge elements that are used for stress analysis, amply demonstrating the performance available in a modern strain gauge.

SURVEY

The term 'strain gauge' is now generally accepted as meaning an electrical resistance strain gauge unless otherwise qualified, but in its widest meaning it embraces all devices which can measure strain and these have made use of all the electrical, mechanical, and optical properties known. Only a few are still used in stress analysis and these are described briefly. The emphasis of this chapter is however on electrical resistance (ER) gauges also commonly referred to as resistance strain gauges, as these account for well over 90 % of the gauges used worldwide. The majority are now metal foil.

The Vibrating Wire (VW) or 'Sonic' Strain Gauge

The VW gauge (Fig. 1) is used extensively in civil engineering, because it has exceptionally good long term zero stability, essential in locations where the load cannot be removed to check the zero. Good stability for periods of twenty years or more is claimed for gauges embedded in the concrete foundations of buildings, dams, and power stations.[1]

The gauge consists of a length of piano wire stretched between two anchor points which are rigidly attached to the material under test. The wire is plucked, and caused to vibrate at its natural frequency, by an electromagnet which is then used as a pick-up to monitor the frequency. The strain is measured as a function of the change in frequency as the tension in the wire is increased or relaxed.

The length of the gauge is usually of the order of four to six inches (100 to 150 mm) although some as small as two inches (50 mm) are available. Embedded gauges are sometimes made up into three-dimensional rosettes of up to nine gauges.[2] Some VW gauges are designed for surface mounting but with their size and bulk they are only suitable for fairly massive structures.[3]

Civil engineering installations very often involve long leads from the gauges to the recording instruments, and the measurement of frequency rather than resistance change is another advantage because errors introduced by the cable lengths are minimal.

Instrumentation can be portable, usually of the 'comparator' type which

FIG. 1. A vibrating wire strain gauge based on an NCB design. Courtesy of Strainstall Limited.

compares the frequency of the gauge against a known frequency to obtain a null. Measuring period rather than frequency yields a greater accuracy and an electronic period counter, with facilities for energising the gauge, forms the basis of some measuring systems. The measurement appears in digital form as the period in microseconds of either one hundred or one thousand cycles of wire vibration. Automatic monitoring of a large number of gauges is achieved with a multi-channel data logging console, which energises the gauges in sequence and records on a printer or punched tape at pre-set time intervals. A more recent development is an instrument which allows dynamic strain measurements to be made from VW gauges up to a strain frequency of 15 Hz.

The Capacitance Strain Gauge

Developed primarily to meet the requirements of measuring static strains at elevated temperature, especially in the 300° to 600 °C range, with acceptable stability, the capacitance gauge is widely used in the electrical

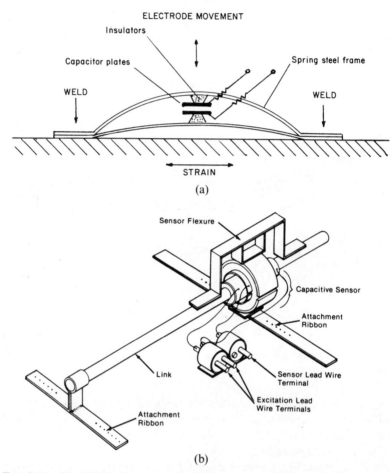

FIG. 2.(a)   The CERL Planer capacitance strain gauge. Courtesy of G. V. Planer Limited. (b) The Boeing capacitance strain gauge. Courtesy of Hitec Corporation.

generating industry in this temperature range. The basic principle utilises the change in capacitance between two plates, the gap between them being altered by the applied strain. Various mechanical means of achieving this can be used and two designs which are available commercially are shown in Fig. 2. The CERL Planer design[4−6] is used extensively in the UK in high temperature measurements and has good long term stability, one of the prime objectives during its development. The Boeing design does not claim such good stability, but is probably better under thermal transient conditions, for which it was primarily developed.[7,8] These two and a third design by Hughes[9] are compared by Sharpe.[10]

The Piezo Electric Gauge
This is the only gauge which is self-generating. It utilises a crystal across which a voltage is developed when a stress is applied. It is used only for monitoring dynamic stresses and is commonly used in accelerometers.

The Piezo Resistive or Semi-Conductor Gauge
The development of the semi-conductor gauge was hailed as the end of the ER foil or wire gauge. The sensitivity was improved by a factor of up to 100 and therefore the demands on amplifiers, which at that time were expensive, were reduced. However unwanted characteristics such as resistance-temperature coefficient, non-linearity, and strain sensitivity/temperature relationships were also magnified in a similar ratio. Perhaps more importantly, the technology which produced the semi-conductor gauge, also made possible the inexpensive stable d.c. amplifier. Now the semi-conductor gauge has won itself an established niche for applications where stiffness of the component or structure is a major requirement, and therefore strain levels are very low. It is also widely used in transducers where a high output is required from a small physical size, and for some dynamic strain measurements where zero stability is not critical.

Modern gauges generally have the strain sensitivity reduced by 'doping' the silicon crystal to give more controllable temperature characteristics.

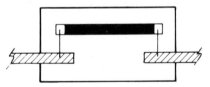

FIG. 3.   A semi-conductor strain gauge. Courtesy of Kulite Sensors Limited.

Typical specifications for a silicon gauge (Fig. 3) are a gauge factor of 100 to 175 (each gauge being individually calibrated); resistance 120 or 350 ohms; linearity within 1 % up to 1000 microstrain; breaking strain approximately 5000 microstrain. Silicon gauges can be produced with either positive gauge factor (P-type) or negative gauge factor (N-type). Because the gauge factor and temperature coefficient depend on the strain level, the initial strain in the gauge after bonding can be significant. Further, because the resistance changes are so much higher than for foil gauges, the non-linearity of the Wheatstone bridge is shown up. All these errors can be corrected for, but it is easier to do so in a transducer, or some device which can be calibrated and compensated before being put to use.[11-15]

The Electrical Resistance (ER) Gauge

This is now by far the most widely used type of strain gauge. Originally made from fine resistance wire on a paper backing, the majority in use today are etched from foils on a synthetic backing such as epoxy or polyimide

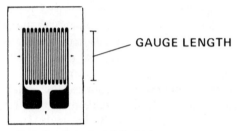

FIG. 4.   A typical foil strain gauge.

(Fig. 4). The many varieties of ER gauges between them cover the temperature range $-270\,°C$ to $+350\,°C$ and, for some applications, as high as $+900\,°C$. Strain range is from a resolution of better than 0.1 microstrain to an elongation of 20% (200 000 microstrain). Gauge sizes range from an active length of 0.008 inches ($0.2\,mm$) to 6 inches ($150\,mm$).

Modern adhesives include 'instant' setting cyanoacrylates for normal temperature stress analysis, a range of cold and hot curing synthetic resins for more demanding applications, and ceramic adhesives for high temperatures. There are also strain gauges which are welded to the test piece either to improve high temperature performance, or to give a hermetically sealed installation for a high hostile environment, or simply to make installation 'on-site' more convenient and faster.

RESISTANCE STRAIN GAUGES—STATE OF THE ART

It is generally accepted that, except for a few very specific applications, foil strain gauges can out-perform wire gauges. They have lower creep under load, better thermal response, better power dissipation, lower transverse sensitivity, and lower scatter in characteristics between gauges of the same type and between gauges in the same batch. For use in precision transducers and weighing scales, the characteristics of a foil gauge pattern can, by small changes in geometry, be modified to match the characteristics of the spring element to optimise performance of the complete unit. Except where specifically stated, all the following data and performance characteristics refer to foil gauges. Information on basic strain gauge theory, design, and installation can be found in the literature[16-20] and in manufacturers' technical data sheets.

Materials
Foil
Copper–Nickel 55–45 alloy (trade names: Advance; constantan) is still the most widely used alloy for both foil and wire gauges, but quality control on the melt, on the thickness tolerance on foils down below 0·0001 inches (0·0025 mm) thick, and the development of techniques for adjusting the resistance temperature coefficient of the foil to make gauges self-temperature-compensated (STC)[21] have allowed giant strides to be made in gauge performance. The strain sensitivity is linear over the elastic range and hysteresis is negligible in a thin foil gauge. In its usual STC form the strain range is up to $\pm 5\%$ but in its super-annealed form, the same alloy can be used for post yield measurements as high as 20% elongation (200 000 microstrain). This alloy is normally used in the temperature range 0 °C to 100 °C but it is often used at higher temperatures in full-bridge installations for limited duration (Fig. 5).

Nickel–Chrome alloys are the next in popularity. There are several different alloys, the most common being Karma (Ni, Cr, Fe, Cu) and modifications of it, which have the advantage of a controllable temperature coefficient for making STC gauges. Modified Karma can be used over a wider temperature range than constantan because it oxidises less rapidly[22] and self-temperature-compensation is better over a wider temperature range (− 270° to + 300 °C) although not as good as Advance close to room temperature.[21] It is chosen in preference to Advance at room temperature when very long term stability is essential.[23,24] It also has the useful property of a negative temperature coefficient of gauge factor which gives some degree of modulus compensation. The coefficient can be adjusted to match specific materials for transducer volume production. Modified Karma also has a better fatigue life than constantan alloy.[15]

At the upper end of the temperature range the STC is destroyed and the resistance changes with time. Sometimes a 'stabilised' form is used instead of the STC form, or Nichrome V is used. However all nickel–chrome alloys

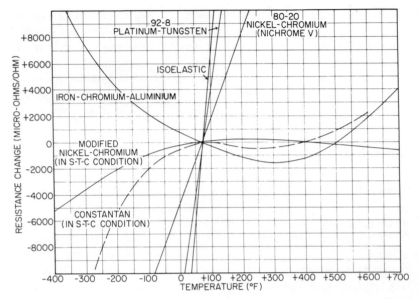

FIG. 5. Temperature induced resistance change for several strain gauge alloys
bonded to steel.

undergo metallurgical order/disorder transformation in the region of 350°
to 550°C accompanied by large changes in resistance making them
unsuitable for measuring static strains. Above this region they can be used
for dynamic measurements or for very short term static measurements.[26,27]

Platinum–Tungsten (8% tungsten) was developed as a strain gauge alloy
because of the problems with nickel–chrome at high temperature. It does
not undergo any metallurgical change up to 900°C and it can therefore be
used for tests involving temperature cycling through the 300° to 600°C
band, and for static strains.[26,27] Platinum–tungsten has a high temperature
coefficient of resistance and cannot be made self-compensating, so great
care must be taken with other compensating techniques when
measurements of small strains are required over a range of temperatures.
The gauge factor of platinum–tungsten is 4 compared with 2 for constantan
and Karma, and it is therefore used sometimes in applications requiring a
higher output which cannot be achieved in other ways, particularly for
dynamic measurements where a high fatigue life is necessary. Strain levels
are usually kept moderate because above 0·3%, the output is non-linear.

Iso-elastic has been the most commonly used alloy for dynamic strain
measurements for many years because it has very good fatigue life[25]

FIG. 6.   A strain gauge installation on a driving wheel of the advanced passenger train. Courtesy of British Rail and Loughborough Consultants.

coupled with a gauge factor of about 3·3. These properties have also made it attractive for some high output transducers. It cannot be self-compensated and it has a high temperature coefficient of resistance. It is normally used in the 0 °C to 200 °C region. Output is non-linear above about 0·5 % strain.

   Manganin is not used for strain gauges, but is used for making surface pressure sensors for measuring pressure waves in the kilobar range, for example for measuring blast waves.[28]

Backing (Matrix) Material
Since the early wire gauges were developed with a paper matrix many synthetic materials have been tried. Those currently in use include epoxies, phenolics, polyesters and polyimides (Fig. 6).

   Polyimides are the most recent and perhaps the most widely favoured for general stress analysis, and for average performance transducers, in the range 0 to 100 °C. Polyimide film is stable, flexible, and extremely tough. It can be used for strain levels as high as 20 %, and is usually treated by the gauge manufacturer to make it compatible with all chemical setting strain gauge adhesives.

Cast epoxies have been in use longer than polyimides but some special versions are still superior for very high performance transducers (i.e. better than 0·5 % spec.). These resins are very hard and brittle making them more difficult to handle and impractical for general stress analysis, but their creep and long term stability are very good.

Epoxy-resin-impregnated fibreglass was originally developed to improve the upper temperature ranges of unfilled resins, but is now used from −270 °C right through to the limit of the organic resins, in the region of +300 °C. It has also been found to give excellent performance in high precision transducers and for long term stability at normal ambient temperatures[23,24] and higher temperatures.[22]

It is worth noting that the traditional tests for peel strength between foil and matrix, or between matrix and specimen have been proved irrelevant and invalid as indicators of performance. The bond strength of an installed gauge used to be tested by attempting to peel the gauge off the specimen with a razor blade which was not possible with older backing materials properly installed. However the polyimide films are so tough that gauges can often be removed in one piece even when properly bonded. In any case it is the shear strength not the peel strength which is important in strain transmission.

Construction
A conventional foil gauge consists of the foil grid etched from rolled foil of about 0·0001 to 0·0002 inches thick (0·0025 to 0·0050 mm), or, in the case of platinum–tungsten it is die-cut because of the difficulties of etching. This is laminated to the matrix which may be any of the materials mentioned earlier, to make an open-faced gauge. The majority of constantan alloy foil gauges are used in this form, but some have an encapsulating layer of the matrix material laminated on top of the foil, sandwiching it and protecting it. The gauge may have short lead wires or ribbons incorporated, or solder dots which protrude through the encapsulation. In some cases the gauge tabs are copper plated or the gauge has integral printed circuit copper terminal strips incorporated to make the attachment of lead wires more convenient and robust.

Karma alloy gauges are usually manufactured with an epoxy-glass matrix, and encapsulation with integral lead-out ribbons or solder dots is fairly standard because the alloy is more difficult to solder by the user.

Weldable strain gauges are available in three basic forms. The first (Fig.

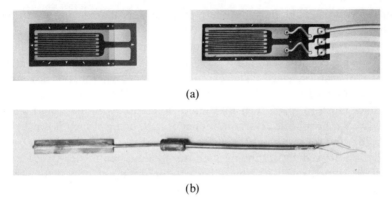

(a)

(b)

FIG. 7. Weldable strain gauges: (a) Micro Measurements (magnified ×2·4), (b) Ailtech (four-fifths actual size).

7a) is simply a standard foil gauge bonded to a stainless steel shim which can then be welded to the test piece 'on-site'. It has the advantage of ease of installation requiring no adhesives on-site, coupled with moderate cost.[29] Working on the same principle, but intended for high temperatures beyond the range of organic materials, the next type has free-filament gauges bonded to the shim with ceramic adhesives.[30] The third type (Fig. 7b) of weldable gauge consists of a small diameter stainless steel tube with flanges down each side which are welded to the test specimen. The tube contains a single loop wire strain element, packed in magnesium oxide to provide both the strain transmission medium and the electrical insulation. This type of gauge can be used for operating temperatures up to 500 °C but also, because the tube is easily sealed, and steel-sheathed cable can be swaged to it to provide an hermetic seal, it is often specified for use under water, or in corrosive atmospheres and other hostile environments where temperature is not the major problem.[31]

Other high temperature gauges of the transferable grid, 'free-filament' or 'strippable backing' variety can be installed directly on the specimen with ceramic adhesives of the phosphoric acid activated type. The installation technique is not easy, and strain levels must be fairly low to ensure good performance and reasonable life. On large specimens achieving the necessary curing temperatures and cycles can be very difficult or impossible.

Installation
Surface Preparation
For most modern organic strain gauge adhesives, a smooth matt finish is

preferred to a rough finish. Modern practice aims at achieving a very thin glue line of the order of 0·0001 to 0·001 inches (0·0025 to 0·025 mm) thick and this cannot be achieved with a rough surface. Wet lapping with silicon carbide paper is the commonly used method, using 400 grit on soft alloys and 280 grit on steels. The exception is for high (post yield) strain measurement. The special adhesives for strains of 10 to 20 % require a much coarser finish to reach their maximum elongation without bond failure.

A weak phosphoric acid 'conditioner' is used on most metals for the wet lapping, and for scrubbing the surface to leave the surface chemically clean. This is followed by scrubbing with an alkaline neutraliser to leave the surface compatible with the adhesives for optimum bond strength. It should be noted that some strain gauge adhesives will not bond satisfactorily to an acidic surface. Recommended surface treatments for a very wide range of materials including most metals and plastics, concrete and wood are given in reference 32. The surface preparation required for weldable gauges is minimal, a hand grinder to remove scale, rust, paint and lumps usually being adequate.

Adhesives

The importance of the adhesive cannot be over emphasised. Its purpose is to faithfully transmit the strain from the specimen into the strain gauge. If it does not do this, the strain gauge readings are meaningless. For this reason *very very few* of the thousands of adhesives commercially available, are satisfactory as strain gauge adhesives, and even these are inferior to adhesives specially formulated or selected by strain gauge manufacturers to get the best performance out of their gauges. It is false economy to use a cheap adhesive from the local hardware shop, or to try to 'discover' a suitable one. The gauge manufacturers are constantly searching for new and better or more universal or easier to use adhesives and those recommended by them should always be used. So often a strain gauge is evaluated by a user (and quite often published as a paper) and the difference between the performance measured, and that specified by the manufacturer is entirely due to the adhesive, which was one the experimenter happened to have available, or one he had used for the past twenty years. The older adhesives gave glue line thicknesses in the range 0·001 to 0·005 inches (0·025 to 0·125 mm) compared with 0·0001 to 0·0002 inches (0·0025 to 0·0050 mm) for some modern epoxies or cyanoacrylates. The reduction in hysteresis and creep under load using the thinner glue lines is quite remarkable, and the improvement in thermal stability, temperature compensation and response to transient parameters is readily seen.

Cyanoacrylates. These are 'instant' adhesives which provide the most convenient adhesive for general stress analysis especially for installation and testing indoors. There are many brands and varieties available, but not all are suitable as strain gauge adhesives (Fig. 8). The better gauge manufacturers sample many batches produced by the adhesive manufacturers and buy only those batches which meet their very demanding

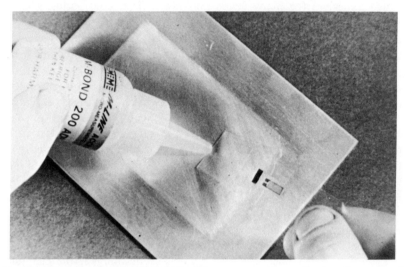

FIG. 8.   Strain gauge being installed with a cyanoacrylate adhesive. Courtesy of Micro Measurements.

specifications. As the gauge manufacturers now have difficulty in finding enough 'good' batches to meet the demand, it follows that *most* of the adhesive sold by other than strain gauge suppliers is *not* satisfactory for strain gauge use.

Cyanoacrylates should always be used with the catalyst supplied by the gauge manufacturer to ensure reliable and repeatable installations. When properly installed using the surface preparation techniques already outlined, the performance is excellent with negligible hysteresis, little creep under load, and capable of elongation to at least 6 % (60 000 microstrain) without bond failure. The limitations are imposed by temperature and humidity. The range of $-5\,°C$ to $+65\,°C$ is much narrower than that recommended by the adhesive manufacturers but is realistic for reliable performance. The adhesive is sensitive to moisture and should *always* have a moisture barrier applied over it immediately after the gauge installation is complete. Because of this moisture sensitivity, the adhesive is not recommended for long term gauge installations. Its life depends on how effectively moisture is kept out. It has been known to last for many months or years, even in difficult environments, but if moisture gets in, bond failure can occur within a few weeks.

Epoxies and epoxy-phenolics. Epoxy cements have been used as strain gauge adhesives for many years, and range from cold curing general purpose adhesives, used mostly for stress analysis at or near ambient temperatures, to very high performance heat curing adhesives used in precision transducers. These latter adhesives are often modified with phenolics.

Generally a heat curing epoxy will always give a better performance than a room temperature curing epoxy, and this is why a wide range is available. The user chooses the convenience/performance compromise, best suited to his requirements. Heat curing epoxies are generally used if testing at elevated temperatures, and the adhesive should be cured and post-cured a few degrees above the required test temperature if best first-cycle results are required in the test. The post cure is necessary as it helps to stress-relieve the installation.

It is important with all epoxies to apply sufficient clamping pressure during curing, to achieve the very thin glue lines which give modern adhesives so much of their performance. With the very high performance heat curing epoxy-phenolics, which are solvent thinned, the clamping pressure also assists in the removal of solvents and entrapped air and so prevents voids under the gauge which would cause gauge instability. Some high-temperature epoxies and epoxy-phenolics can operate for limited periods at over 300 °C. This is the limit for organic materials which break down by oxidation and sublimation.

Polyesters. The fast curing of some polyesters at low temperatures makes them very attractive for applications where speed is important or heating cannot be applied. However performance is inferior to epoxies in most respects and they are mostly used where performance can be sacrificed for convenience, or where there is no alternative because of ambient conditions during installation. The operating temperature range for cold curing polyesters is up to 150 °C. Some are capable of giving good performance on the first temperature cycle after cold curing, but their most attractive property is the ability to cure at temperatures as low as 4 °C whilst retaining a usable pot life at 24 °C. It should be noted that polyester adhesives are not always compatible with all types of strain gauge matrix materials, especially some epoxies.

Phenolics. Phenolic resins have largely been superseded by epoxies and epoxy-phenolics which are less demanding in clamping pressures and curing cycles, and give better performance.

Polyimides have a good potential as high-temperature adhesives, and a few are available commercially. So far they have proved difficult to use successfully with strain gauges because of a tendency to bubble under the gauge, because of entrapped solvents.

Ceramic adhesives. Organic materials can rarely operate for any length of time above 300 °C, therefore ceramic adhesives, and 'free-filament' or 'strippable' gauges, i.e. without a matrix, are used. The installation procedure is lengthy, and requires a skilful operator. The failure rate is high unless the operator is good and has frequent practice. The cements are based on aluminium phosphate and silicon in a water-based slurry which is air dried and then baked. Curing will usually be completed 10 to 20 degrees above the required operating temperature. The maximum operating temperature is 700 °C long term, or 800 °C short term. If the build-up of cement is too thick it may crack or even flake off the surface. This is definitely a technique which should only be used when there is no alternative.

Flame-sprayed alumina. This is not really an adhesive, but it is a means of bonding a free filament gauge to a test item for high temperature testing. It is used extensively for testing 'hot' components in gas turbines and has proved satisfactory for some dynamic measurements up to 1000 °C. Static performance is limited by the gauge alloy performance at high temperatures rather than by the alumina.[33] The special equipment and expertise required have prevented its wider use, and it is most suitable for installing gauges on components which can be transported to a special spray booth (Fig. 9).

The technique is to spray droplets of molten alumina onto the surface of the test piece to provide an insulating layer, and then attach the free filament gauge to this by spraying a further layer, or layers, embedding the gauge in the alumina. The special guns are developed from those used to spray metals, and use alumina in either rod or powder form, feeding it to an oxy-acetylene flame where it is melted into droplets which are propelled to the test piece by the gas plasma. Wire gauges are invariably used with this technique because the droplets flow more easily under and around the fine circular cross-section wire and give a better bond with less likelihood of voids under the gauges. The technique was originally developed using guns employing alumina rods, and these are still used, but powder guns have been found better for installing very small gauges, that is below $\frac{1}{8}$ inch (3 mm) gauge length.

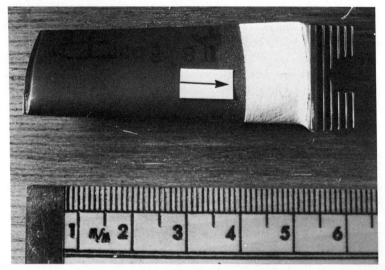

FIG. 9. High temperature gauge installed on a turbine blade using flame-sprayed alumina. Installation by Hitec Corporation.

Weldable gauges. In weldable gauges, the primary objective is to remove the difficulties inherent in cementing a gauge 'on-site'. Initially, the emphasis was on high temperature gauges where the curing requirements for high temperature adhesives often cannot be met, especially on a large structure. By cementing a gauge to a thin stainless steel shim under laboratory conditions, and attaching this to the component on-site by spot welding, the curing limitations are removed. More recently, weldable gauges have been used increasingly at normal temperatures, simply to ease the installation problems on-site, and in the case of hermetically sealed weldable gauges, for operating in hostile environments, such as under the sea or in solvents or corrosive gases. With the portable battery operated welders now available installation is very quick and convenient.

Lead Wire Attachment and Selection
Some foil strain gauges, and all wire gauges, are supplied by the manufacturer with short lead wires or ribbons already attached, and the user solders or welds his lead wires to these. However, the majority of foil gauges are supplied, and used, without pre-attached leads, sometimes for cost reasons, but primarily because in general it is much easier and more convenient to bond gauges properly and with uniformly thin glue lines, if they do not have

leads attached. Methods of attaching leads and some recent developments to simplify the procedure are discussed briefly, but reference should be made to instructions and technical literature supplied by the gauge manufacturers. For most applications over the temperature range $-270°$ to $+260°C$ soft solder is perfectly satisfactory provided that the alloy chosen as suitable for the temperature. For a general purpose solder over the normal ambient temperature range, a good quality 63/37 tin–lead

FIG. 10. A standard gauge installation for stress analysis.

eutectic alloy is preferred for wetting and flow characteristics. It should be in fine wire form, with a core of *non-corrosive* rosin-type flux. Many fluxes which are claimed to be non-corrosive are sufficiently corrosive to significantly change the resistance of a strain gauge over a period of time. Even the fluxes which are acceptable for strain gauges must be removed completely with a suitable solvent because they will affect the resistance to ground, and the stability, of a completed gauge installation. For cryogenic temperatures, the solder should be inhibited against 'tin disease' by the addition of antimony, or a low (5·2 %) tin alloy is used. This latter alloy can also be used at higher temperatures, up to 260 °C. Above approximately 260 °C and up to 650 °C there is the choice of silver soldering or welding. Silver solder requires a resistance soldering tool, and welding requires a special welder designed specifically for welding fine wires or ribbons to each other. Above 650 °C there is no alternative to welding.

Most strain gauge users now prefer to use printed circuit terminal strips (Fig. 10), bonded to the test piece close to the gauges, to take the lead wire proper, and fine jumper wires from the terminal strip to the gauge tabs. A jumper wire may be a separate length of fine gauge solid-conductor copper

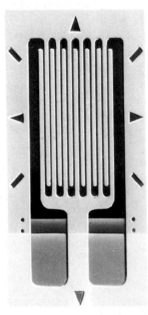

FIG. 11.   Robust encapsulated gauge suitable for direct attachment of stranded
instrument wire. Courtesy of Micro Measurements.

wire, or one of the strands from the lead wire continued through the
terminal strip to the gauge. This technique reduces the risk of damage to the
gauge should the wires be accidentally snatched or pulled.[34] The
temperature limitations imposed by the terminal strips should be noted.
Foil gauges have been developed with integral terminal strips to reduce
installation time, and more recently, encapsulated gauges with specially
treated tabs, which can take stranded instrument wire soldered directly on
the gauge tabs, have proved very popular and convenient for installation
under less-than-ideal conditions (Fig. 11). The choice of lead wires is
dependent on the environment, the temperature range, and the length of
lead required. Copper is the obvious first choice usually tinned and
stranded, with the heavier gauges used for longer lead lengths to keep the
resistance within manageable limits. Polyvinyl chloride (PVC) is the most
widely used insulation, but the quality can vary considerably between
manufacturers. Lower quality PVC wire is sometimes porous. It is only
suitable in the laboratory. Better quality material is perfectly satisfactory
for outdoor use and under water, even under pressure. It is, however, wise to

test all wires to be used under water by coiling them in a bucket of water and checking insulation to ground. It only takes one pin-hole to ruin a test!

Outside the PVC temperature range, there is a choice of polytetrafluoro-ethylene (PTFE) or polyimide insulation. PTFE is the more common and it can be surface etched so that protective coatings and adhesives will stick to it satisfactorily. Like PVC the quality can vary depending on the manufacturer. The good quality material is satisfactory for under-water installations, and the temperature range is $-260\,°C$ to $+260\,°C$. For the higher temperatures the copper should be silver-plated. Polyimide insulation is extremely tough and abrasion resistant and is chosen for this reason as often as for its higher temperature limit of $315\,°C$. A disadvantage is that it is not readily available in colour coded form. All the above types of wire are available in three-conductor and four-conductor flat cable, which is very convenient for strain gauge connections. There are, however, applications where flat cable should *not* be used, examples being in electrically 'noisy' areas where shielding is required, and in strong magnetic fields where tightly twisted cable is needed to minimise the 'loop' area.

Nickel-clad copper can be used up to $480\,°C$ and it is available as a solid conductor wire insulated with fibreglass. Fine nickel–chrome ribbon ($980\,°C$ maximum) without insulation is used for short connections welded to strippable backed gauges. Because of its high resistivity, lengths are kept to a minimum, changing to nickel-clad copper wire immediately outside the high temperature zone. For heavy duty or permanent installations, mineral insulated cable with copper or stainless steel sheath gives completely sealed watertight cable runs which are very resistant to mechanical damage.

Gauge Protection

The long term protection of resistance strain gauges continues to be a problem and a hermetically sealed metallic enclosure is the only practicable method for water immersion. However, for more general short term requirements an adequate range of coating materials is offered by strain gauge manufacturers. Some of the more popular classes are mentioned here but the suppliers' literature should be consulted for performance limits. Commercially available coating materials from other sources should not be used without thorough evaluation as many will react with the foil or solder of an open-faced gauge producing an unstable installation. Examples of these are cold-curing silicone rubbers which have acetic acid active in the curing reaction. They are known to be very good waterproofing materials, but they will corrode a strain gauge very rapidly.

Brush-on coatings—air drying. These are usually solvent-thinned synthetic materials which dry by evaporation in a short period of time at normal temperatures. Several are available including polyurethanes (Fig. 12), RTV silicone rubbers, acrylics, and silicone varnishes. They provide a very convenient coating method which is perfectly adequate for many installations indoors, or outdoors when protected from the worst of the weather. Some of them, especially acrylics, are used as primer coats on open-faced gauges before applying more robust protection.

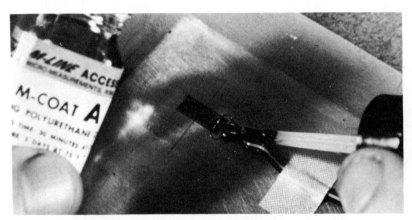

FIG. 12.   A gauge protected with a polyurethane coating.

Brush-on coatings—hot curing. Usually used for operating temperatures above 90 °C, many of the strain gauge adhesives are included in this group of coatings. Epoxies, polyimides and silicone varnishes are the most common.

Two-component chemical setting. Epoxy coatings, usually rubberised or highly plasticised and often thixotropic, fall into this category; very commonly used for protection from water immersion up to about 70 °C and for pressures up to several kilobars for limited periods. As no synthetic materials are completely impervious to moisture, the 'life' of the coating under water is proportional to the effective thicknesses which can be applied and the cleanliness of the substrate and inversely proportional to the degree of stability required from the gauge installation.

Air curing silicone rubbers (RTV). There are many room-temperature curing (RTV) silicone rubbers available, but as already mentioned those which vulcanise by an acetic acid reaction should not be put into direct contact with strain gauges. There are others, recommended by gauge suppliers, which can be used safely. They are a convenient alternative to the two-part epoxies, and often used for short-term under-water protection. They are rather less impervious than the epoxies, take longer to cure, but being less stiff cause less reinforcement of the component.

Butyl rubber sheet. This very sticky mastic is supplied in sheet form and is simply pressed over the gauge installation in the form of a patch. It sticks tenaciously, and requires no air drying, or curing time making it very suitable for use 'on-site'. It is often given mechanical protection by pressing

FIG. 13.   Gauge protected with butyl rubber, neoprene rubber, and aluminium foil. Courtesy of Micro Measurements.

a sheet of neoprene rubber on the outside and is sometimes covered also with self-adhesive aluminium foil to give a smooth contour (Fig. 13). Care must be taken with the gauge lead wires coming out of the protected area to avoid moisture ingress around them. Whilst the butyl rubber is an excellent moisture barrier, its electrical resistivity can vary. It should not therefore be in direct contact with an open-faced gauge, bare solder dots or wires.

Lead wire sealing. The problem area in a strain gauge protection system is often the interface between the lead wire insulation and the protective coatings. PVC (Vinyl) leads should have a primer coat of nitrile rubber solution which will give a good͵bond to most synthetic coating materials. PTFE (Teflon) wires should be etched with one of the proprietary etching solutions, otherwise no coating will bond to it satisfactorily. Some wires can be purchased already etched. Polythene wires are seldom used in this sort of installation because of the difficulty in sticking to them.

Under-water protection. A typical protection system for use under water (Fig. 14) to give a life of a few weeks would be:

1.   The surface of the specimen should be thoroughly cleaned round

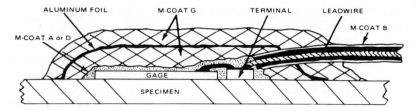

FIG. 14.   Cross-sectional view of typical long term installation. Courtesy of David Taylor Model Basin. Some of the earliest work on the incorporation of metal barrier layers in strain gauge protective systems was done by Mills Dean III at the David Taylor Model Basin. A further description of this technique may be found in *Strain Gage Waterproofing Methods and Installation of Gages on Propellor Strut of* USS Saratoga, SESA, Oct. 1957.

      the gauged area, as for installing a gauge. The width of this band should ideally be at least $2\frac{1}{2}$ cm if space permits. ,

2.   A coat of acrylic or polyurethane lacquer brushed over the gauge, solder joints, and any bare lead wire, extending about 3 mm into the cleaned area.

3.   A thin coat of nitrile rubber solution applied to the PVC wire insulation extending about 8 cm from the gauge. This should be allowed to dry.

4.   A coat of thixotropic epoxy or RTV silicone rubber applied 2 to 3 mm thick over the gauge area, one centimetre into the cleaned area, and 4 cm along the lead wires. *Note:* if the coating used does not have enough mechanical strength to hold the lead wire securely, then the lead wire must be anchored firmly to the specimen close to the protected area, to prevent movement of the wire.

5.   When the first coat has dried, a second similar coating should be applied.

6.   Sometimes a layer of aluminium foil is incorporated between the two layers of epoxy, to provide an impermeable barrier. In salt-water an outer coating of nitrile rubber may also be helpful.

For longer term under-water protection, a complete metal enclosure is necessary. This can be made from stainless steel or aluminium, with a flange which can be spot-welded or cemented to the specimen, and with a 'nozzle' for the cable exit. The whole enclosure should be filled with a suitable mastic or wax and should be used *in addition* to the protection already described. Alternatively proprietary mineral-filled cables with standard glands and connectors can be used with a more substantial metal enclosure over the gauge area. The mineral-filled weldable gauge with integral lead wires in stainless steel sheathing can be used without further protection. A technique of vulcanising rubber patches over strain gauges on ships' propellers has recently been used successfully in the USA.[35]

INSTRUMENTATION

Most strain gauge instrumentation utilises a Wheatstone bridge or some modification of it, the only exception being the use of potentiometric circuits for some purely dynamic testing.[17,36]

Static Strain
Digital displays have largely replaced analogue in static strain indicators, improving resolution without sacrificing range, and instrument stability has been improved to match. Commercial strain gauge instrumentation still offers the choice of A.C. carrier or D.C. but the latter has gained ground with the improvements in D.C. amplifiers in recent years. Null balance systems are still popular, especially for portable battery operated instruments, because stabilised voltage supplies are not required. Most manufacturers have recognised the importance of power dissipation in small gauges or on non-metallic specimens and offer a variable bridge energisation or a fixed voltage which is low enough for the majority of requirements.

Data logging has made rapid strides and many systems now incorporate micro-processors to carry out signal conditioning and data reduction (Fig. 15). Some have on-line computers for rapid data analysis or storage. These facilities have so far been used mainly in the large systems with 100 or more channels, but the trend is for much smaller and less expensive loggers to have micro-processors and soft-ware instead of multi-channel signal conditioning hardware. Some static strain indicators using D.C. amplifiers have the facility for dynamic recording on pen recorders or galvanometers, and some have a 'peak hold' facility for monitoring single shot dynamic strains.

Dynamic Strain (Fig. 16)
Multi-channel strain gauge amplifiers, including facilities for bridge completion, bridge balance, variable bridge energisation, amplifier zero and amplifier gain usually have at least one calibration resistor built in.

FIG. 15. A strain gauge data logging system employing a microprocessor and software for signal conditioning. Courtesy of Intercole Systems Limited.

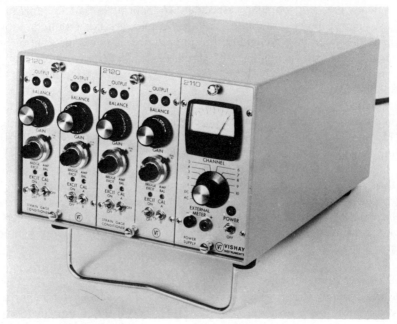

FIG. 16. Multi-channel amplifier/signal conditioning system for dynamic strain measurement. Courtesy of Vishay Instruments.

They are used in conjunction with pen recorders, galvanometer or magnetic-tape recorders. Cathode-ray oscilloscopes are often most convenient for displaying the outputs of one or two channels. Both A.C. carrier and D.C. systems are available, but D.C. has gained ground in popularity. The frequency response for carrier systems is usually below 10 kHz and for D.C. amplifiers 0 to 25 kHz. For high frequency strain D.C. systems are available for up to 100 kHz. Some amplifiers have built-in filters to reduce noise problems outside the frequency range being studied, and some systems offer a choice of constant voltage or constant current gauge energisation.

PERFORMANCE LIMITS FOR ELECTRICAL RESISTANCE STRAIN GAUGES

The parameters affecting strain gauge performance are so many and interdependent that it is almost impossible to quote performance figures without several pages of qualifications and constraints. Some years ago, one of the most important of these was probably the name of the experimenter! Current installation techniques, adhesives and gauge quality control allow almost anyone to achieve a reliable and repeatable standard of gauge installation after a little practice, even outside the 'normal ambient' temperature range.

Maximum temperature is a typical example where it is virtually impossible to quote a meaningful number. If good stability is required for a period of months, the temperature limit is certainly below 300 °C but what is meant by good stability? To some it may mean tens of microstrain to others hundreds, to some ± 5 microstrain in which case the limit will be below 200 °C and perhaps below 100 °C depending on other factors. For purely dynamic strains there have been successful measurements at over 900 °C. For static strains above 300 °C the capacitance gauge should be considered.

Long term stability conversely is very dependent on the temperatures involved. At room temperature, gauges have been evaluated for monitoring dimensional changes in metal samples over long periods and have shown stability within a few parts per million.[23,24] For higher temperatures gauges have been evaluated for creep testing at 200 °C for 500 h.[22]

Resolution of a properly installed gauge is very much better than 0·1 microstrain but the usefulness of this is limited by the stability of the installation over the time scale and temperature range of interest.

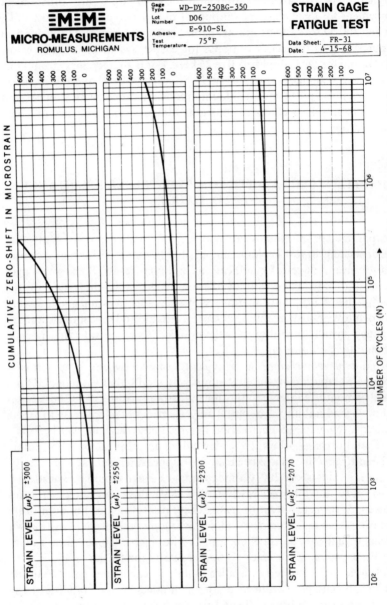

Fig. 17. Fatigue life of a $\frac{1}{4}$ inch (6mm) gauge made from iso-elastic alloy. Courtesy of Micro Measurements.

Repeatability is of the same order, and is often better than the repeatability of the material being tested.

Hysteresis is so low that it is seldom mentioned. Precision transducer manufacturers find that in developing load cells to better than 0·01% accuracy, the steels used have to be very carefully selected and heat treated to make them as good as the strain gauges.

Insulation to ground is a good check on the quality of a gauge installation. If long-term stability is required the insulation should be very high and should remain so. If a gauge is installed under normal clean dry conditions, it should not be difficult to regularly exceed 20 k megohms. If 2 k megohms is not being achieved then there is something not being done correctly, probably flux residue is being left on the gauge tabs. However for the gauges installed on-site an insulation as low as 2 megohms might have to be acceptable and will probably be adequate for the accuracy required under those conditions.

At the bottom of the scale, a ceramic adhesive installation will give an insulation resistance as low as 2 k ohms at room temperature but this will improve on heating as the moisture is driven out. It will fall again above 700 °C.

Fatigue life is defined by different manufacturers in different ways. Some quote cycles to open circuit, i.e. complete failure, others quote cycles to a defined shift of gauge zero, for example 100 microstrain, which occurs long before complete failure.[25,37]

Fatigue life depends not only on gauge materials (Fig. 17) but on size, lead wire attachment, uniformity of strain over the gauge area, and of course on the strain level. A typical life for a constantan gauge of $\frac{1}{4}$ inch (6 mm) gauge length with a fully reversed cyclic strain of ± 1500 microstrain is 10^6 cycles. For a $\frac{1}{4}$ inch (6 mm) gauge in iso-elastic alloy with epoxy glass backing, a typical life for ± 2500 is 10^7 cycles.

Maximum elongation. Most standard constantan gauges, in self-temperature-compensated form, will stand a single excursion to 3%, i.e. 30 000 microstrain. If the gauge length is $\frac{1}{4}$ inch (6 mm) or more this is usually improved to 5%. However in its super annealed form, constantan gauges, over $\frac{1}{4}$ inch (6 mm) gauge length, will go to 20%, i.e. 200 000 microstrain. It should be noted however that this is for a single excursion, and open-faced gauges are preferred. It should also be remembered that work-hardening, accompanied by a change in resistivity, will occur if the super annealed foil is strain cycled.

A common problem is that a gauge fails at an indicated strain lower than the limit expected. This is usually because there is a severe strain gradient under the gauge, causing the gauge to fail locally whilst indicating the _average_ strain under the gauge area.

Cryogenic temperatures. Gauges and adhesives have been found which are satisfactory for temperatures down to $-270\,°C$.[38]

Magnetic fields. Some errors are introduced by strong magnetic fields. These include: magnetic pickup by the wiring, which can be minimised by using tightly twisted wire, and by shielding; magnetic pickup in the gauge, which can be eliminated by using special double gauges non-inductively coupled; magneto resistive effects, which are lower in Karma foil than in constantan; and magnetostrictive effects.[39]

Centrifugal forces on rotating components of over $100\,000\,g$ are regularly encountered. The gauges are seldom a problem as they have very low mass and a large bonding area. The lead wires need very secure anchorage and glass cloth tape, impregnated with an epoxy adhesive and bonded to the surface, is often used for this purpose. Ideally it should be wrapped around the component to hold the wiring in place. Sometimes the wiring can be routed to the inside face such as the inner face of a flange on a turbine disc to minimise the problem. The attachment of lead wires to the gauges can be critical under high-g conditions. The size and mass of solder joints should be kept to a minimum. Fine wires should be used for the connections and a layer of adhesive impregnated glass cloth cemented over the completed installation.[40]

Hydrostatic pressure is not a problem, unless the pressurising medium is water or some fluid which attacks the protective coatings used. It is essential to use a very thin glue line under the gauge completely free from voids, and a flexible protective coating is usually preferred. The effect of hydrostatic pressure has been investigated by several researchers, at pressures as high as $400\,000\,psi$ (27 kilobars)[41,42] and the errors shown to be less than -7 microstrain per thousand psi (70 microstrain per kilobar), on a quarter bridge.

Recent Applications and Techniques
Residual Stress Measurement by the Hole Drilling Method[43-45]
This technique has gained in popularity because it can often be considered non-destructive, particularly on large components and structures. The hole, either $\frac{1}{8}$ or $\frac{1}{16}$ inch (3·75 or 1·588 mm) diameter and depth the same, can easily be drilled on-site and recordings made on a portable strain indicator. Special rosettes are available for this technique which have the hole position accurately located relative to the three gauges, which are arranged round the hole in a $45°/90°$ rosette configuration. Drilling rigs

FIG. 18. Alignment and drilling jig for residual stress measurements. Courtesy of Photolastic Inc.

incorporating optical microscopes for accurately aligning the cutter axis with the hole centre available commercially. For cutting particularly hard materials, an airbrasive cutter can be used instead of an end mill, and this equipment is also available commercially (Fig. 18).

Strain Measurement on Composites
Foil strain gauges are being used successfully on glass reinforced plastics, carbon fibre composites and many others. Special care needs to be taken over power dissipation[46] and local reinforcement of the specimen by the gauge may need to be corrected for. There are very few published references available to-date.

Strain Measurement on Non-reinforced Plastics
The problems of power dissipation and local reinforcement are very much greater than on composites. The local heating from a gauge which has too high a voltage across it not only causes thermal drift problems, it can also reduce the modulus of the plastic locally or introduce thermal stresses in the model or component. However, many researchers are working on a wide

TABLE 1

ALLOWABLE GAUGE POWER DISSIPATION IN $WATTS/INCH^2$ ($WATTS/MM^2$) FOR A PROPERLY INSTALLED FOIL STRAIN GAUGE. COURTESY OF MICRO MEASUREMENTS

Accuracy Requirements		EXCELLENT Heavy Aluminum or Copper Specimen	GOOD Thick Steel	FAIR Thin Stainless Steel or Titanium	POOR Filled Plastic such as Fiberglass/Epoxy	VERY POOR Unfilled Plastic such as Acrylic or Polystyrene
STATIC	HIGH	2.–5.0 $(3.1 \times 10^{-3} - 7.8 \times 10^{-3})$	1.–2. $(1.6 \times 10^{-3} - 3.1 \times 10^{-3})$	0.5–1. $(0.78 \times 10^{-3} - 1.6 \times 10^{-3})$	0.1–0.2 $(0.16 \times 10^{-3} - 0.31 \times 10^{-3})$	0.01–0.02 $(0.016 \times 10^{-3} - 0.031 \times 10^{-3})$
	MOD.	5.–10. $(7.8 \times 10^{-3} - 16 \times 10^{-3})$	2.–5. $(3.1 \times 10^{-3} - 7.8 \times 10^{-3})$	1.–2. $(1.6 \times 10^{-3} - 3.1 \times 10^{-3})$	0.2–0.5 $(0.31 \times 10^{-3} - 0.78 \times 10^{-3})$	0.02–0.05 $(0.031 \times 10^{-3} - 0.078 \times 10^{-3})$
	LOW	10.–20. $(16 \times 10^{-3} - 31 \times 10^{-3})$	5.–10. $(7.8 \times 10^{-3} - 16 \times 10^{-3})$	2.–5. $(3.1 \times 10^{-3} - 7.8 \times 10^{-3})$	0.5–1. $(0.78 \times 10^{-3} - 1.6 \times 10^{-3})$	0.05–0.1 $(0.078 \times 10^{-3} - 0.16 \times 10^{-3})$
DYNAMIC	HIGH	5.–10. $(7.8 \times 10^{-3} - 16 \times 10^{-3})$	5.–10. $(7.8 \times 10^{-3} - 16 \times 10^{-3})$	2.–5. $(3.1 \times 10^{-3} - 7.8 \times 10^{-3})$	0.5–1. $(0.78 \times 10^{-3} - 1.6 \times 10^{-3})$	0.01–0.05 $(0.016 \times 10^{-3} - 0.078 \times 10^{-3})$
	MOD.	10.–20. $(16 \times 10^{-3} - 31 \times 10^{-3})$	10.–20. $(16 \times 10^{-3} - 31 \times 10^{-3})$	5.–10. $(7.8 \times 10^{-3} - 16 \times 10^{-3})$	1.–2. $(1.6 \times 10^{-3} - 3.1 \times 10^{-3})$	0.05–0.2 $(0.078 \times 10^{-3} - 0.31 \times 10^{-3})$
	LOW	20.–50. $(31 \times 10^{-3} - 78 \times 10^{-3})$	20.–50. $(31 \times 10^{-3} - 78 \times 10^{-3})$	10.–20. $(16 \times 10^{-3} - 31 \times^, 10^{-3})$	2.–5. $(3.1 \times 10^{-3} - 7.8 \times 10^{-3})$	0.2–0.5 $(0.31 \times 10^{-3} - 0.78 \times 10^{-3})$

variety of plastics, with perhaps Perspex models being the most numerous. Satisfactory results are being obtained by using very low power levels, sometimes using pulsed energisation, and by correcting for the reinforcing effect of the gauge[47,48] (Table 1).

Analysis
Computer programs are available for the analysis of the data from rosette gauges when large numbers of gauges are involved. Some gauge manufacturers can also provide formulas for the temperature characteristics of gauges which can be used in computer analysis.[49,36] For small numbers of gauges or for those without access to a computer, programs are also available for at least one of the popular portable programmable calculators.

The standard equations or nomograms for rosette analysis are available in the literature[16,20] and discussions on the errors in rosette analysis caused by transverse sensitivity have been published.[50,51] It is worth noting however that the largest errors are often in the assumed values of Young's modulus and Poisson's ratio for the material being tested. The analysis of dynamic and vibrational stress is a more complex subject and techniques are often developed to meet particular requirements.[52]

Services
In the past ten years there has been a rapid growth in the service industries.

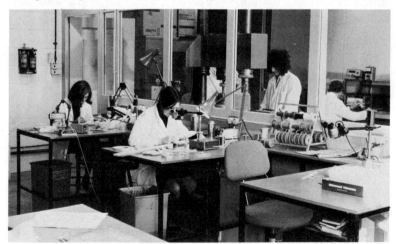

FIG. 19. Strain gauge installation services. Courtesy of Welwyn Strain Measurement Limited.

These now provide anything from training in basic strain gauge installation techniques to a complete installation, recording, and data analysis service either in the laboratory (Fig. 19) or on-site anywhere in the world. Many companies with occasional strain measurement problems on components now find it more economical or quicker to send the components to a specialist company to have the gauges installed, wired and checked out.

CONCLUSION

There are very few practical strain measurement problems which cannot be tackled with strain gauges, and a search of the literature will almost certainly find someone who has already tried or done something similar or relevant.

In some cases where there are alternatives or preferred techniques, for example photoelasticity or moiré, a familiarity with strain gauge methods and an awareness of the errors involved, coupled with the advantages of recording continuously or remotely, can often outweigh the apparent advantages of the other techniques. Carrying out a survey with a brittle coating followed by detailed point measurements with strain gauges is a very powerful combination of techniques which should always be considered when the locations of stress concentrations or gradients, or the directions of principal stresses, are not known.

REFERENCES

1. HORNBY, I. W., *Strain*, 1972, **8**, 110.
2. BABUT, R. and BRANDT, A. M., *Strain*, 1977, **13**, 18.
3. PARROTT, L. J., *Strain*, 1973, **9**, 146.
4. NOLTINGK, B. E., MCLACHLAN, D. F. A., OWEN, C. K. V. and O'NEILL, P. C., *Proc. I.E.E.*, 1972, **119**, 897.
5. NOLTINGK, B. E., *Experimental Mechanics*, 1974, **14**, 420.
6. PROCTOR, E. and STRONG, J. T., Transducer '77 Conference, 1977, Trident International Exhibitions Ltd, Tavistock.
7. HARTING, D. R., 21*st Int. Instr. Symp.*, Instr. Soc. Am., 1975, ASI 75251, 289.
8. CORUM, J. R. and SARTORY, W. K., ASME, 1973, 73–WA/PVP–4.
9. GILLETTE, O. L., *Experimental Mechanics*, 1975, **15**, 316.
10. SHARPE, W. N., *Experimental Mechanics*, 1975, **15**, 482.
11. WALLACE, R. H., *Strain*, 1972, **8**, 162.

12. BAKER, M. A., *BSSM Meeting*, Edinburgh, February 1978. Reprint Kulite Sensors Limited.
13. DEAN, M., (ed.) *Semi-conductor and Conventional Strain Gauges*, Academic Press, New York, 1962.
14. *Manual on Experimental Stress Analysis*, 2nd edn., Chapter 2, SESA, 1965.
15. DORSEY, J., Strain measurement with semi-conductor strain gauges, *I.S.A. Conf.*, Los Angeles, 1965.
16. PERRY, C. C. and LISSNER, H. R., *The Strain Gauge Primer*, 2nd edn., McGraw-Hill, New York, 1962.
17. PROCTOR, E., *Methods and Practice for Stress and Strain Measurement*, Allison, I. M. (ed.), BSSM Newcastle, 1977.
18. HEARNE, E. J., *Strain Gauges*, Merrow Publishing, Watford, 1971.
19. WINDOW, A. L., *An Introduction to Strain Gauges*, 5th edn., Welwyn Strain Measurement Ltd, 1978.
20. VAUGHAN, J., *Strain Measurements*, Bruel & Kjaer, Naerum, Denmark, 1975.
21. *Strain Gauge Temperature Effects* 1976, TN–128–2 Micro Measurements, Michigan, USA.
22. FAIRBAIRN, J., *Strain*, 1974, **10**, 175.
23. MARSCHALL, C. W. and HELD, P. R., *Strain*, 1977, **13**, 13.
24. FREYNICK, H. S. and DITTBENNER, G. R., *Experimental Mechanics*, 1976, **16**, 155.
25. *Fatigue of Strain Gauges* 1974, TN–130–3 Micro Measurements, Michigan, USA.
26. EASTERLING, K., *Brit. J. Appl. Phys.*, 1963, **4**, 79.
27. BERTODO, R., *J. Strain Analysis*, 1965, **1**, 11.
28. DUGGIN, B. W. and BUTLER, R. I., 1970, ISA Conference Paper 623.
29. *Weldable Strain Gauges and Temperature Sensors*, 1976, PB–112–2 Micro Measurements, Michigan, USA.
30. DAVID, T. J., *Strain*, 1968, **4**, 4.
31. GIBBS, J. P., *Experimental Mechanics*, 1967, **7**, 19.
32. *Surface Preparation for Strain Gauge Bonding*, 1976, B–129 Micro Measurements, Michigan, USA.
33. WNUK, S. R., Development of strain gauges for vibratory strain measurements on advanced gas turbines, *Proc. TCFSG, SESA*, 1972.
34. *The Proper Use of Bondable Terminals*, 1975, TT–127–4 Micro Measurements, Michigan, USA.
35. DEAN, M., *Experimental Mechanics*, 1977, **17**, 303.
36. BEANEY, E. M., *Methods and Practice for Stress and Strain Measurement* Allison, I. M. (ed.), BSSM Newcastle, 1977.
37. DOWLING, N. E., *Experimental Mechanics*, 1977, **17**, 193.
38. TELINDE, J. C., *Experimental Mechanics*, 1970, **10**, 394.
39. WALSTROM, P. L., *Cryogenics*, May 1975, 270.
40. *Techniques for Bonding Lead Wires to Surfaces Experiencing High Centrifugal Forces*, 1975, TT–132 Micro Measurements, Michigan, USA.
41. MILLIGAN, R. V., *Experimental Mechanics*, 1964, **4**, 25.
42. FOSTER, C. G., *Experimental Mechanics*, 1977, **17**, 26.
43. RENDLER, N. J. and Vigness, I., *Proc. SESA*, 1966, **XXIII**, 2, 577.
44. BEANEY, E. M. and Proctor, E., *Strain*, 1974, **10**, 7.

45. BEANEY, E. M., *Strain*, 1976, **12,** 99.
46. *Optimising Strain Gauge Excitation Levels*, 1977, TN–127–3 Micro Measurements, Michigan, USA.
47. EPELLE, O. B., *Strain*, 1975, **11,** 17.
48. BERME, N., MENGI, Y. and TARHAN, A., *Strain*, 1975, **11,** 169.
49. MORETON, D. N. and MOFFAT, D. G., *Strain*, 1975, **11,** 66.
50. MEYER, M. L., *Experimental Mechanics*, 1967, **7,** 476.
51. *Errors Due to Transverse Sensitivity in Strain Gauges*, TN–137 Micro Measurements, Michigan, USA.
52. PEARSON, D. S., *Strain*, 1976, **12,** 24.

Chapter 5

DEVELOPMENTS IN MOIRÉ AND LASER METHODS OF STRESS ANALYSIS

A. R. LUXMOORE

University College, Swansea, UK

SUMMARY

Moiré and laser methods of stress analysis are large field, non-contact methods which have always required specialised knowledge and equipment. Their optical principles are very similar, because they use the processes of diffraction and interference between light rays to measure the deformation of solid bodies. Recent developments have been aimed at making these techniques more robust without sacrificing their inherent sensitivity, thus making them suitable for field work.

This chapter describes the most significant developments in these techniques that have occurred during the past decade, and compares their sensitivities.

INTRODUCTION

The determination of stresses by experimental methods often requires the measurement of several physical parameters. The most common requirement is for in-plane strain measurements on the surfaces of a component, but in certain circumstances sufficient information can be obtained by measurement of out-of-plane displacements or changes in slope and curvature of the surfaces, as, for example, in plate bending problems. Moiré and laser measurements can be used for all these parameters, and a brief review of the variety of applications of the moiré effect has been given previously by the writer[1] and also by Theocaris[2] and Durelli and Parks.[3]

161

Moiré and laser methods are grouped together in this chapter because of their similarity both in optical principle and also in application. Both are large field, non-contact techniques requiring rather specialised expertise and laboratory conditions. Most recent work has been aimed at extending these techniques to industrial environments and practical strain measurement.

MOIRÉ STRAIN MEASUREMENT

Basic Principles

Figure 1 illustrates the basic principles of the moiré effect, produced by the overlapping of lines and spaces of regular gratings due to either a difference

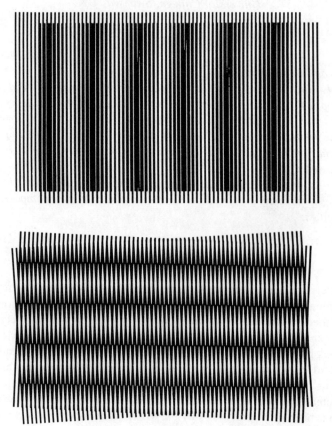

FIG. 1.   Principle of moiré effect.

in pitch or an angular misalignment. In a more general case, the two effects are combined, so that the fringes cross the grating lines at some angle. If the pitch and angular misalignment vary across one grating, the fringes will also be curved.

For in-plane strain measurement, a grating is deposited on to the specimen surface, either by bonding, as with conventional strain gauges, or by photoengraving. A transparent reference (or master) grating is

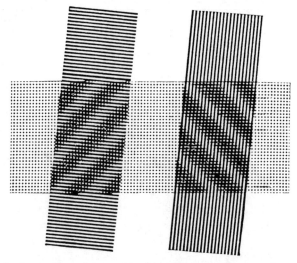

FIG. 2.   Orthogonal specimen grid with master gratings superimposed at right angles.

superimposed either in contact or in the image plane of a projection system. As the specimen is strained, moiré fringes will be formed due to changes of pitch and angle in the specimen grating. It is easily shown[1] that the fringes represent contours of in-plane displacement, as measured perpendicularly to the grating lines. For in-plane deformation, two orthogonal displacement parameters, usually designated u and v, are necessary, and the strain parameters are related to these by:

$$\varepsilon_x = \partial u/\partial x \qquad \varepsilon_y = \partial v/\partial y \qquad \gamma_{xy} = \partial u/\partial y + \partial v/\partial x \qquad (1)$$

Hence two orthogonal gratings are needed on the specimen for a complete strain evaluation, and these may be deposited simultaneously, producing an orthogonal array of dots, Fig. 2. To avoid confusion between the two

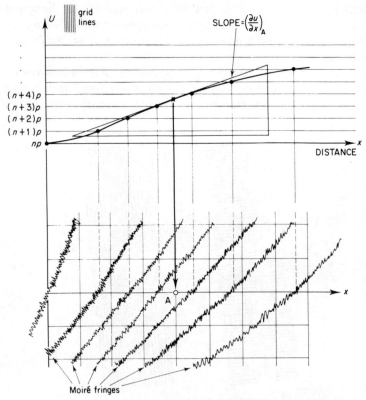

FIG. 3.   Analysis of strain at points of intersection of the network using displacement–distance plot.

fringe patterns (referred to as u- and v- isothetics), the master grating remains a line grating, aligned first with one specimen grating, then rotated through exactly $90°$ to be aligned with the other grating. This procedure ignores the specimen grating currently at $90°$ to the master grating.

From Fig. 1 and eqn. (1), it can be seen that the normal strain ε_x (or ε_y) can be obtained by measuring fringe spacings perpendicular to the grating lines, and the component of shear $\partial u/\partial y$ (or $\partial v/\partial x$) by measuring parallel to the grating lines. For a varying strain field, it is usual to plot the fringe order (equal to some multiple of the grating pitch) against distance along an orthogonal network of lines parallel and perpendicular to the grating lines, Fig. 3, and determine the derivatives graphically at the points of intersection of the network. This is done on identical networks for both specimen gratings and then the complete strain information is obtained at each intersection of the network.

Sensitivity

The main disadvantage of the conventional moiré technique outlined above is lack of sensitivity. Gratings are available commercially for moiré strain measurement with densities of up to 40 lines/mm (1000 lines/in). Strains can be measured from direct fringe patterns using these gratings with a sensitivity of around 1000 microstrain, which is outside the working strain range of many common structural materials. Suitable interpolation techniques can increase this sensitivity by an order of magnitude.

The most common interpolation procedure is the initial mismatch between master and specimen grating. This can be an angular misalignment, a difference in pitch, or a combination of the two. The mismatch is constant over the grating area, producing parallel fringes of equal separation. Changes in the specimen grating will distort the initial fringe pattern, and analysis can be carried out by subtracting the apparent initial displacements from the final displacements.

This procedure also simplifies the determination of tensile and compressive strain, for if an initial pattern is produced by making the master pitch larger than the specimen pitch, a tensile strain will increase the fringe spacing, and vice versa for a compression. In practice, where the sign of the strain is known in advance, it is always best to arrange the mismatch so that there is an increase in the number of fringes.[4]

With a mismatch, the sensitivity depends on the accuracy with which the fringe centres can be located. Experience has shown that, even with electro-optical methods of fringe location, this is at best accurate to 1 % of the fringe spacing. Hence the displacement sensitivity is no more than 1 % of the pitch, and the strain sensitivity will depend on the effective gauge length over which the displacement is measured. If this is large, e.g. a constant strain field over the area of the specimen grid, then the strain sensitivity is usually well within the elastic range ($< 0.1\%$) of most materials.

A mismatch between master and specimen grating extends the range of the moiré technique by producing fringe patterns in strain fields where no fringes would occur with the direct moiré effect. This is very useful for constant strain fields, but does not help with plotting strain gradients. In fact, a large mismatch will often obscure a small strain gradient: the mismatch will produce closely spaced fringes in which small variations of fringe spacings will be ignored.

The plotting of strain gradients can be improved by including the positions of fractional fringe orders (cf. photoelasticity). These can be obtained by moving the master grating a fraction of a pitch perpendicular to the grating lines. If, for example, this fraction is one half pitch, then the fringes produced correspond to displacements of $(n + \frac{1}{2})p$, $(n + 1\frac{1}{2})p$, $(n + 2\frac{1}{2})p$, etc., as shown in Fig. 4. The subdivision may be finer than this, of course, but there is no point in plotting fractions smaller than 0·1 pitch, as

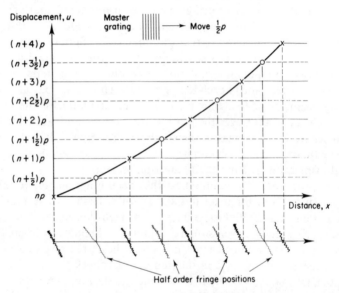

FIG. 4.   Plotting half order fringe orders by moving master grating one half pitch.

the uncertainty in determining the fringe centre will approach this value. Also movement of the master grating must be very precise.

This procedure for plotting fractional fringe orders does not increase the basic sensitivity of the moiré technique, as does the mismatch technique, and these two interpolation methods are complementary rather than alternatives. They are essential for elastic strain evaluations.

Attempts to improve moiré sensitivity have been aimed at increasing the line density of the specimen grating. Line densities greater than 100 lines/mm are not generally available for moiré work because of the difficulty of reproducing these gratings photographically,[9] and even when they are available, formation of moiré fringes requires special spectroscopic systems. The sensitivity of existing gratings can be improved by utilising their diffraction effects or using improved interpolation methods. Alternatively, very fine gratings (up to 3000 lines/mm) can be produced interferometrically, with the help of a laser.

Diffraction Effects

A regular grating will diffract incident light into discrete directions,[5] and these directions can be identified by the spectra observed in the back focal plane of a decollimating lens, Fig. 5(a). Each direction is designated as a diffraction order, with the undiffracted light numbered zero. As the order increases, the amount of light diffracted reduces rapidly, and it is quite

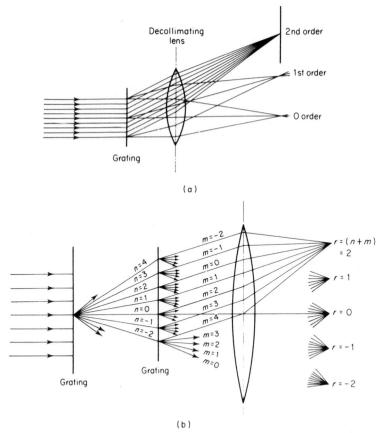

(a)

(b)

FIG. 5.   Diffraction effects of (a) single grating showing stopping out of second order, and (b) superimposed gratings.

usual for moiré gratings to diffract useful amounts of light only into the first order. Although the higher orders (harmonics) can be obliterated, the two first orders must always be present if the basic grating structure is to be reproduced. In general, the presence of the higher orders indicates a grating with sharply defined lines and spacings, and if such a grating is to be imaged precisely, all these diffraction orders must be transmitted by the lens. All lenses are diffraction limited, and this limit can be reduced further by the effect of lens aberrations. (In Fig. 5, the diffraction limit of the decollimating lens is equivalent to an aperture placed in the back focal plane, effectively stopping the high order spectra from being transmitted.) Physically, it is more accurate to consider the contrast of a grating image to be gradually reduced as the lens aperture is reduced until all details of the grating is lost when the first diffraction order is excluded.

When two gratings with similar pitches are superimposed, Fig. 5(b), the diffracted orders of the first grating are diffracted again by the second, but because the diffraction angles are the same for both, the diffracted spectra appear the same as for a single grating, although each spectral spot now contains several orders. The full analysis of this system of multiple diffraction is very complex,[6] but, in general, if all orders except the nth is stopped out, then the number of fringes observed is n times the number observed in an unmodified moiré system, i.e. the sensitivity is increased n times. Unfortunately, the fringes are not of equal contrast,[3] and this can cause considerable errors when estimating the fringe centre. In practice, this simple arrangement is not particularly useful, but Post[7] modified it by using a master grating with a line density equal to n times the specimen line density. Using the first order position of the master grating, he effectively ignored all orders except the nth order from the specimen grating, and this produced a simple two-beam interference effect with fringes of equal contrast, but a sensitivity equal to n times the specimen grating line density. Of course, this requires the specimen grating to diffract some light into the nth order, which places a considerable restriction on the preparation of this grating. However, as will be shown later, there are certain types of grating where this is readily achieved.

This process of 'optical filtering' can also be achieved when imaging one grating onto another via a lens,[3,8] but the specimen grating must still diffract light into the appropriate order, as before.

Interpolation Methods
The writer has developed two methods of interpolation which can be used to measure small strain gradients. They do not increase the basic sensitivity

over that obtained using a conventional mismatch, but they provide more information than is available by the conventional fractional fringe shift method described earlier and hence allow more accurate determination of the gradients.

The first was developed for situations where both large tensile and compressive strains are present, e.g. bending, thus making a single mismatch insensitive to some area of the specimen grating (where the mismatch approached the specimen strain). In many instances, it is convenient to fix the master grating to the specimen, and photograph the fringes directly under load, so that only one mismatch can be used, as in Fig. 6(a). In such circumstances, it is preferable to use a large rotational mismatch, Fig. 6(b), and first plot the displacement pattern parallel to the grating lines, i.e. along lines A, B, C, D, etc., in Fig. 6(b). These lines can be quite close together, but the fringe positions along each line must be plotted in the correct relation to the other lines, Fig. 6(c). This diagram now represents the overall displacement picture, and the displacements perpendicular to the grating lines can be obtained by measuring along lines I, II, III, etc. on Fig. 6(c), and plotting the displacement versus distance in the usual way to obtain the tangents, Fig. 6(d).

The second method is really an extension of fringe shifting as discussed previously and requires the specimen (or a photograph of it) to be moved by precise amounts perpendicular to the grating lines. The principle is illustrated in Fig. 7, where an unstrained specimen grating is displayed under a small mismatched reference grating containing a fiducial mark. If the displacement $x = np$, where n is an integer and p is the specimen pitch, a fringe will still be located under the fiducial mark. If the specimen grating is strained uniformly over the distance x, producing a grating pitch p', the specimen grating must now be moved an amount $x + \partial x = np'$ for a fringe to appear under the fiducial mark again. Hence the strain is given by

$$\varepsilon = \frac{p' - p}{p} = \frac{\partial x}{x}$$

The advantage of the method arises from the very small values of x which can be chosen when plotting strain gradients. A further important advantage is that location of fringe centres by movement of the fringes is far more accurate than measurement of the static pattern because the background noise does not affect the appearance of the moving fringes. The method can be used to determine all three strain parameters, but if only one parameter is to be measured along a single line, care must be taken to ensure

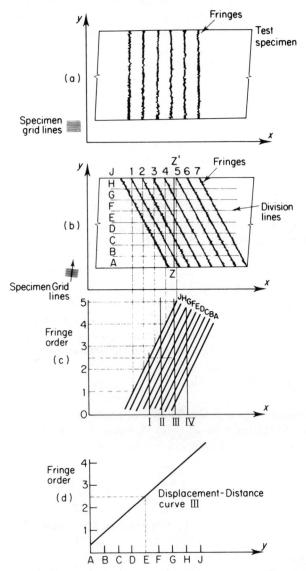

FIG. 6. Interpolation of moiré fringe pattern using rotational mismatch; (a) mismatch fringe pattern, (b) loaded fringe pattern, (c) displacement–distance curves parallel to grid lines, (d) displacement–distance curve perpendicular to grid line.

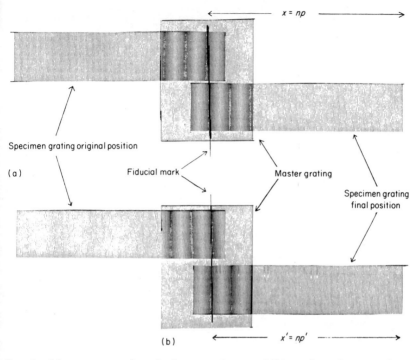

FIG. 7.   Measurement of strain from continuous shifting of specimen grating across the fiducial mark on the master grating; (a) before straining specimen, (b) after straining.

the specimen grating is displaced along the same line before and after straining.

These two interpolation methods were used to measure the transverse strains in a beam subjected to both direct stress and bending moment. Comparison with the conventional technique, Fig. 8, shows the improvement that is obtained.

Interferometric Gratings
The significant scattering of light by very fine gratings (> 100 lines/mm) creates problems in both reproducing the gratings and observing the resulting moiré fringes by placing two gratings in contact. It can be shown[9] that gratings reproduce images of themselves at distances of approximately np^2/λ, where n is an integer and λ is the wavelength of the illuminating beam. As p becomes very small, this repeat distance becomes so small that all the

172   A. R. LUXMOORE

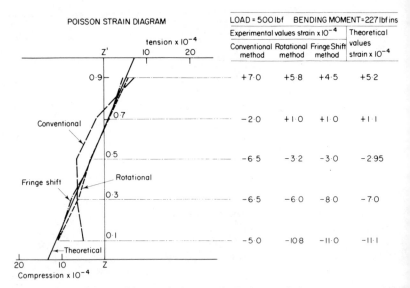

POISSON STRAIN DIAGRAM			LOAD = 500 lbf BENDING MOMENT = 227 lbf ins			
			Experimental values strain x 10^{-4}			Theoretical
			Conventional method	Rotational method	Fringe Shift method	values strain x 10^{-4}
0·9			+7·0	+5·8	+4·5	+5·2
0·7			−2·0	+1·0	+1·0	+1·1
0·5			−6·5	−3·2	−3·0	−2·95
0·3			−6·5	−6·0	−8·0	−7·0
0·1			−5·0	−10·8	−11·0	−11·1

FIG. 8. Comparison of interpolation methods for analysing transverse strains across a beam.

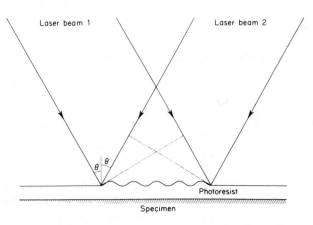

Mean wavelength of corrugations equals mean fringe spacing = $\lambda / 2 \sin \theta$.

FIG. 9. Production of high density moiré phase grating on positive photoresist by interference of two coherent laser beams.

diffraction orders become incoherent, and the effective depth of field of the grating is negligible. This problem can be avoided by optically selecting a single diffraction order, as described previously, and this is the method used to observe moiré fringes in spectroscopic optical systems. Alternatively, the gratings can be produced by the interference of two coherent beams of light making some angle with each other, Fig. 9, and recording this grating on a photosensitive material. This system has a very large depth of field if two laser beams are used, and the moiré fringe pattern can be formed by interfering the recorded pattern with that produced interferometrically.

Cook[10] used an Argon laser to produce interferometric gratings on non-flat objects coated with a blue sensitive photoresist. The photoresist was a positive emulsion,[11] so that the developer only dissolved away the exposed areas of the emulsion, leaving the grating in the form of approximately sinusoidal surface corrugations. The specimen was distorted, and then re-exposed so that a 'frozen' fringe pattern was formed on the surface (positive photoresists can be exposed and developed several times without detriment). Where the two gratings were completely out of phase, the emulsion surface was left plane, but where they were in phase, the corrugations were reinforced, and these scattered light into discrete orders, thus providing a contrasting fringe pattern when the emulsion was viewed in reflected light.

Wadsworth, Marchant and Billing[12] developed the technique further, and used real-time observation of fringes in their studies on carbon fibre composites, which enabled them to observe behaviour under increasing load. They also had to use Argon laser, but Boone[13] was able to use a much cheaper Helium–Neon laser by bonding red-sensitive holographic film to the specimen, and immersing the whole specimen in the processing tanks. This technique provides a considerable increase in sensitivity over the previous moiré methods, but requires very careful and time-consuming experimentation. In general, if this order of sensitivity is required, it is probably more convenient to use a speckle technique, as discussed later.

Some Practical Techniques
The principles outlined above have been combined to produce several self-contained moiré techniques. The writer has developed rapid grid reproduction techniques[14,15] based on orthogonal grids made of very thin metal sheets which are available commercially with up to 80 lines/mm. These grids can either be bonded directly to the specimen, using strain gauge techniques, or used as stencils, which allows the same grid to be used several times. The specimen grating can be copied using a contact

photographic method,[16,17] which can be adapted for work in ambient light conditions, and the photograph (or the actual specimen) examined on a modified profile projector.[4] These projectors give magnified images from × 5 up to × 100, and the modifications allow continuously variable mismatch, as well as continuous specimen displacement for interpolation purposes. Recently, the writer has adapted one of these instruments to provide fringe multiplication using the specimen grating harmonics, but with photographic copies the harmonics are often of very low light level, and the aberrations in the lenses also limit the resolution of very high orders.

Photographic copies (or transparent specimens) can also be examined in a diffractometer,[18] using a reference grating in near contact. The reference grating can be a photographic copy of the specimen in the unstrained state, and the individual fringe patterns can be separated by using appropriate filtering.[19,20] This is useful when using specimen gratings with large ruling errors in them, as these errors cancel out. For maximum sensitivity, it is best to use a bonded metal grid, and copy it by replication. The writer and his colleagues have recently developed a rapid replicating procedure of high stability and sensitivity,[21,22] and this can be used to replicate the bonded metal grids. The replicas, which are transparent, reproduce the grid harmonics up to the 10th order, producing an effective line density of up to 1600 lines/mm, but these high orders can only be observed in the diffractometer, with two grids in near contact.

A similar system has been developed at the Centre of Industrial Innovation at Strathclyde University. Available commercially under the name of Optecord, the system consists of specially prepared phase gratings with a very sharp, square profile. These are deposited onto the specimen via an epoxy-resin mould, which, once cured, allows the master grating to be removed. This moulded grating retains the sample profile as the master, and subsequent replicas of this mould are taken using a fast-curing silicone rubber backed by a glass plate. The sharpness of the original grating is retained by the replicas, and these can be projected via an optical system to produce up to the 10th order of diffraction. Unfortunately, as the replicas are opaque, the optical system is rather expensive, and necessitates the use of a laser.

Grids can also be recorded with cameras, but apart from the filtering effect of the lens, which diminishes the harmonic content of the recorded grid, this procedure can also produce errors due to out-of-plane movements of the specimen during loading.[19,23] These errors can be corrected either by using a telecentric lens,[19,23,24] or a multi-axis viewing system[23] analogous to photogrammetry. A camera is essential if large scale moiré grids have to be recorded, and Burch and Forno[25,26] have used a modified 35 mm Pentax camera to record moiré gratings up to several metres square. They apodise the lens of the camera so that it will record a grating of 300 lines/mm in the image plane. The corresponding size of the object grating will depend on the magnification used, but they have used specimen gratings ranging from 2 up to 40 lines/mm, whilst retaining the same image line density. The moiré fringes are observed in the direction of the first diffraction order from the processed negative.

HOLOGRAPHIC INTERFEROMETRY

Principles

A simplified explanation of holography is given by Lehman[27] and for the present purpose, it is sufficient to consider the arrangement of Fig. 10. The laser beam is diverged by the beam expander, and part of it strikes the object and part is reflected by the mirror onto the photographic plate to form the reference beam. The light scattered by the object interferes with the reference beam and forms a pattern of standing waves, the intensity of

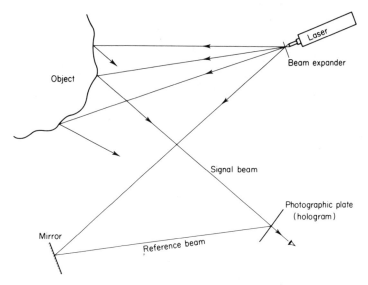

FIG. 10.   Principle of holographic recording.

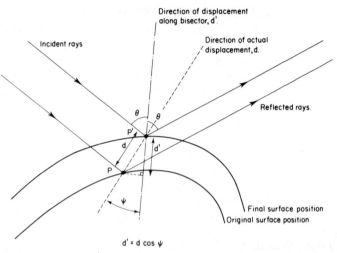

d' = d cos ψ

FIG. 11. Geometry of displacements measured by holographic interferometry.

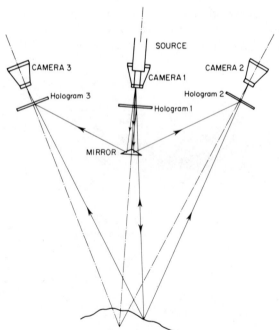

FIG. 12. Determination of absolute displacement from three non-planar hologram positions.

which is recorded by the photographic plate. This plate, after processing, becomes the hologram. If the hologram is replaced in its original position, and illuminated with the *reference beam only* (by covering the object with a black card) the hologram will scatter part of the reference beam as if the light came from the object, i.e. with the same relative amplitudes and phases, so that on looking through the hologram one will perceive the object. If the object is now uncovered, it will also scatter light with the same relative amplitude and phases as produced by the hologram, and hence reinforce the light scattered by the hologram (providing the hologram is pre-positioned in exactly the same place), because the light is coherent, i.e. there is a constant phase relation between any two points along the laser beam. If the object is now moved or distorted, the light scattered by the object will change its relative amplitudes and phases, and interference will occur between the light scattered by the object and that scattered by the hologram. Similarly, a change in the position of the hologram will also cause interference, and the hologram must be repositioned to within a fraction of a micron to avoid errors.

When viewing the fringe pattern 'live', i.e. with the original position of the specimen reconstructed by the hologram and interfering with the direct light from the distorted specimen, the contrast is often poor, because the amplitudes of the two interfering beams are quite different. It is often impaired even further by the presence of small random vibrations caused by the operator leaning over the apparatus to inspect the fringes. These problems can be avoided by recording both undistorted and distorted holograms on the same photographic plate, and reconstructing a 'frozen' fringe pattern.

Haines and Hildebrand[28] analysed the resulting interference pattern for the general case, but Ennos[29] produced a simplified analysis which showed that holographic interferometry measures the displacement along the bisector of the angle between the incident and reflected rays, Fig. 11. This is the main principle used when evaluating holographic interferograms.

Measurement of Total Displacement

In general, the displacement at any point on a surface must be specified by three independent coordinates (usually Cartesian, designated u, v, w), and three independent holographic measurements must be made at each point on the specimen. This can be achieved by illuminating with a single beam, and recording with three separate holograms arranged in some non-planar configurations, Fig. 12. For any point on the surface, the bisectors between incident and reflected light can be determined for each hologram from the

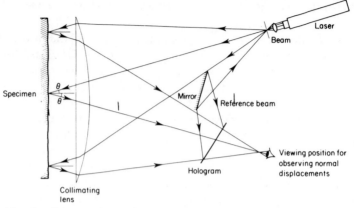

FIG. 13. Possible experimental arrangement for studying out-of-phase displacements of flat surface.

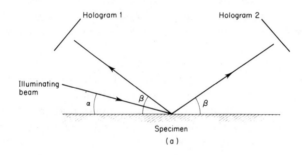

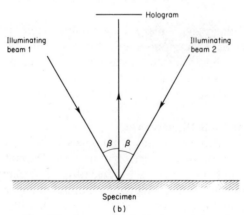

FIG. 14. Measurement of in-plane strain by holograph interferometry using (a) Ennos' two-hologram system, (b) two illuminating beams with single hologram.

geometry of the system, and the displacement along these bisectors determined from the order of interference on each hologram, i.e.

$$n\lambda = 2d_i \cos \theta$$

where n = order of interference fringe, λ = wavelength of light, d_i = displacement resolved along ith bisector, and 2θ = angle between incident and reflected rays.

By determining these three vectors at several points on the surface, the Cartesian displacements can be determined. A computer program has been developed for this purpose.[30] One major difficulty with this process is the determination of the absolute fringe order, which is not known for each hologram. Though the absolute fringe order is not essential, the relative relation between fringe orders on each of the holograms must be known, otherwise their results cannot be correlated. In practice, this means the absolute displacement of some point on the surface must be determined by other means, for example, by keeping one point fixed in space.

An alternative method for determining the complete displacement at any point makes use of the finite aperture of the hologram, and was suggested independently by Gates[31] and Tsujiuchi.[32] A pinhole is placed over the hologram, and a point on the surface is viewed through this hole. Because the fringes are formed at infinity, they take on a distinctive appearance, being straight, parallel fringes for translation parallel to the hologram, and circular if the displacement is away or towards the hologram. In general, the pattern is a mixture of the two, and they can be interpreted by a single graphical method.[33] For pure translation, the displacement can be determined by moving the pinhole across the hologram, and counting the number of fringes crossing the point under observation on the specimen.[31] Neither of the above two methods has proved popular for practical applications, mainly because of the complexities involved in analysis.

Measurements on Flat Surfaces

The measurement of out-of-plane displacements on a flat surface can be achieved fairly readily by using collimated incident illumination and a decollimating lens behind the hologram, both axes of collimation making equal angles to the surface normal, Fig. 13. Light and Luxmoore[34] have used a similar arrangement to detect cracking in concrete cubes and cylinders by observing discontinuities in the fringe patterns, and this approach has proved quite useful in the field of non-destructive testing.[35]

In-plane displacements cannot be determined quite so readily. Ennos[29] attempted this first by taking separate holograms, Fig. 14(a), inclined so that when the fringe orders obtained from each hologram were subtracted, they cancelled the out-of-plane displacement component, and left only the in-plane displacements. Though successful, this procedure was extremely time-consuming in computation, and absolute fringe orders had to be determined for each hologram (by keeping one end of the specimen clamped).

Subsequently this arrangement was simplified by Butters,[36] Luxmoore and House[37] and Boone.[38] They all used a single hologram, with two illuminating beams inclined at equal but opposite angles to the surface normal, Fig. 14(b). This arrangement produced a dual fringe pattern on the hologram which represented both in-plane and out-of-plane displacements. The presence of two superimposed fringe patterns complicated the interpretation of the hologram, and this method has now been replaced by speckle interferometry, which uses the same basic geometric configuration to separate in-plane and out-of-plane displacements.

SPECKLE INTERFEROMETRY

Origin of Speckle

Speckle pattern interferometry utilises the speckled appearance of an optically *rough* surface illuminated by highly coherent illumination, such as a laser beam. The speckles arise because of the microscopic interference effects that occur between the light rays as they are scattered randomly by the surface. This is not observed with ordinary, or incoherent, light because there is no fixed phase relation between light rays coming from different parts of the source, and the would-be speckles are averaged out. The size of the speckles is governed by the wavelength of light used and the aperture of the viewing system, but in most practical cases, it will always be small (< 0.1 mm). There is no limit to the roughness of the surface for producing a speckled appearance (except that it should be fairly random), and so the technique can be applied to most engineering components.

Speckle interferometry can be used to measure three parameters important in stress analysis: in-plane displacements, out-of-plane displacements, and surface tilt. It is a large field technique, capable of covering areas of a few millimetres square up to several metres square. It can be applied to both flat and curved surfaces, but its main advantage over holographic interferometry is the separation of in-plane and out-of-plane displacements on flat surfaces. Leendertz[39] recognised the potential of the speckle phenomena for separating components of displacement, and his original technique has been improved on by many other workers.

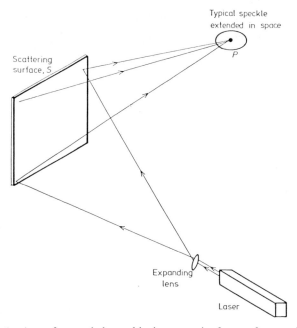

FIG. 15.   Production of extended speckle in space in front of scattering (or diffusing) surface.

Distribution of Speckles

When a rough surface is illuminated with a laser beam, the surface will scatter light in all directions. At any point P in front of the surface S (Fig. 15) the light waves from each scattering point will add vectorially to produce a light intensity which will depend on the size of the scattering surface (diffuser) and the position of P relative to it.[40] This phenomenon manifests itself as an array of speckles distributed in the space in front of the scatterer, and which can be viewed by placing a screen in a suitable position. The speckles are generally ellipsoid in shape and become larger as the scatterer becomes smaller, and as the distance from the scatterer increases. For practical purposes, the size of the scatterer and the distance from it will always produce very small speckles.

The speckle pattern can be viewed directly by a lens system but in this case, the size of the speckle in the image plane of the lens is controlled by the finite aperture of the lens, which collects light over a solid angle controlled by the aperture. If the surface of the scatterer is focused onto a screen in the image plane, there is a one to one relationship between points on the

scatterer and points on the image plane. However, because of the coherent
nature of laser light, speckles will exist everywhere within the volume on the
image side of the lens, and they can be viewed (or photographed) by placing
a screen (or film) in positions other than the image plane. This point has
important consequences in the applications of speckles, as will be seen later.

Formation of Young's Fringes

When a speckle pattern is recorded photographically, using high resolution
film, it is quite common to record the same pattern twice on the same
negative, but with a small displacement between the two patterns, caused by
some distortion of the object. When the film is processed and examined
under a microscope, it is possible to identify pairs of individual speckles.[41]

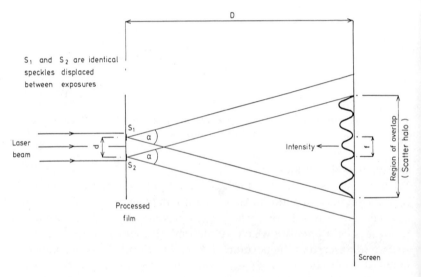

FIG. 16.   Production of Young's fringes from doubly exposed speckle photograph.

More usefully, if a narrow beam of light (not necessarily coherent) is passed
through the negative, both speckle patterns will scatter the light in an
identical manner (due to the identical speckle shapes and distribution), but
with the scattering displaced in the direction of the displaced speckles, Fig.
16. This phenomenon is identical to the classical Young's experiment, where
two closely spaced slits are used to scatter light from a small source, and
produce fringes within the intersection of the scattered cones of light, due to

interference between scattered rays that are coherent *relative* to each other. Similar interference fringes are produced from the two displaced speckle patterns, and for obvious reasons, these are called Young's fringes. The spacing of the fringes are inversely proportional to the displacement between the speckle patterns, and the fringes are aligned perpendicular to the direction of the displacement.

In Fig. 15, the undiffracted beam passing through the negative will obscure the Young's fringes over quite a large area. A lens can be placed just behind the negative (on the same side as the screen) in order to focus the main beam to a point on the screen. In more precise physical terms, the Young's fringes are formed in the back focal plane of the lens, and this pattern represents the Fourier transform of the doubly exposed negative. In this situation, the area of the collimated illuminating beam can be made as large as the replica, providing the displacement is constant over this area. For varying displacements, the beam should cover an area over which the displacement does not vary by more than 10%. The distance, D, over which the cone of light is spread now equals the focal length of the lens.

Fringes of Deformation

Measurements on Young's fringes can only determine the *average* displacement between two speckle patterns, and in most stress analysis problems we are more interested in *variations* in displacements between two patterns, e.g. a constant strain arises from a linear variation in displacement across a surface. These variations can be indicated by optical filtering of the doubly exposed negative as shown in Fig. 17. The negative is illuminated by a parallel beam of light, and imaged onto a screen by a lens. For small variations in displacement, Young's fringes will be seen on a screen placed in the back focal plane of this lens, but for larger variations, the Young's fringes will become progressively more 'washed-out', because different areas of the negative will produce different Young's fringe patterns. If a small aperture is placed off-axis in the back focal plane, Fig. 17, the only light passing through the aperture will be that diffracted in that particular direction, corresponding to bright Young's fringes. This light will produce bright regions on the image plane, separated by dark regions, and this fringe pattern is related to the deformation of the original speckle distribution. The actual relationship will depend on which parameter is being measured, but the deformation fringes usually represents contour intervals of the parameter.

In the writer's opinion, deformation fringes are less useful than Young's fringes because their contrast is much lower. This is because the emulsion

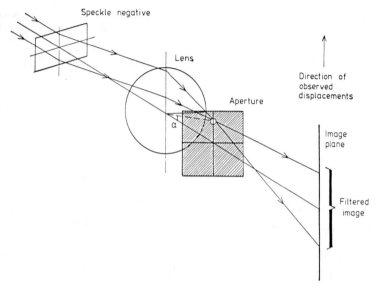

FIG. 17.   Spatial filtering of doubly exposed speckle negatives to produce fringes of
deformation.

and other components of the optical system will scatter light on a random
basis, and this stray light contributes a considerable portion of the total
light passing through the aperture, thus lowering the signal/noise ratio. The
Young's fringes, however, are observed over the whole back focal plane,
and the stray scattered light is far less important.

Measurement of In-Plane Displacement
The original method suggested by Leendertz[39] used the basic arrangement
of Fig. 14(b), but the hologram is replaced by a camera using high
resolution holographic film and focused on the specimen surface. Each
illuminating beam produces its own speckle pattern, but these combine into
a composite speckle pattern in the camera image plane. Small movements
along the z- or y-axes result in identical phase changes for the individual
speckles, causing no relative phase change in the composite pattern, and
hence no intensity change. A movement, d, in the x-direction produces a
phase change of $2\pi d \sin \theta/\lambda$ for one beam and $-2\pi d \sin \theta/\lambda$ for the other
beam, giving a resultant phase change (and hence a corresponding intensity
change) of $4\pi d \sin \theta/\lambda$. Leendertz showed these phase changes by
photographing the composite speckle before deformation, processing the
negative and replacing it in its original position. Black spots on the negative

now covered the bright speckles, and vice versa, so that no light was transmitted by the undistorted object through the negative. When the object was deformed, the relative phase changes due to in-plane displacement in the x-direction changed, and where before there were 'dark speckles', i.e. regions producing zero intensity, some light occurred and is transmitted through the negative. Provided the displacements are small compared with the size of the speckles, the speckle patterns before and after deformation remain correlated, and the negative is covered by bright and dark fringes corresponding to contours of in-plane displacement.

This same basic arrangement was used by Archbold, Burch and Ennos[42] to produce speckle interference fringes from a doubly exposed negative, using the non-linear characteristics of the film. This was quite successful provided displacements were small compared with the speckle size, i.e. the two patterns remained correlated. Once displacements exceeded the speckle size, they observed Young's fringes in the back focal plane of the lens imaging the doubly exposed negative onto a screen. They also found that they could observe fringes of deformation by placing a spatial filter in the position of a dark Young's fringe, and these fringes were equivalent to those observed when using the non-linear film characteristics. Butters and Leendertz[43] explained this phenomenon by considering the speckle pattern as a record of spatial frequencies ranging from zero to the resolution of the recording lens. The change in speckle intensity caused by in-plane displacement is equivalent to a rearrangement of the spatial frequency content so that regions which do not correlate, i.e. have phase changes other than multiples of 2π, will show no Young's fringes, and will scatter light in the direction of the dark Young's fringes, whilst regions having phase changes of 2π will scatter no light.

Duffy[44] modified the two-beam speckle method by using a double aperture in his lens and a single illuminating beam. This is equivalent to viewing the specimen from two equally but oppositely inclined directions, and produces a fine grating pattern within each speckle. This grating changes its position within the speckle as the speckle phase changes (owing to in-plane displacement). With a doubly exposed negative taken before and after deformation, the regular changes in phase of the speckles produces a moiré pattern representing contours of displacement. The technique works only for image displacements smaller than the speckle size, because identical speckles must overlap for the speckle gratings to remain correlated. The sensitivity is the same as that using two-beam illumination, but the angles of observation are limited by the finite aperture of the recording lens.

At this stage, it became apparent that there were two methods[42] available for measuring in-plane displacements; a two-beam (or double aperture) system, insensitive to small out-of-plane movements provided the speckles remained correlated; a single-beam method[45] which produced Young's fringes (and fringes of deformation) related to the in-plane displacements provided these were larger than the speckle size. It was suggested[45] that the first method was suitable for measuring displacements of several wavelengths of light, i.e. a few micrometres, whilst the second method could measure displacements of several speckle diameters, i.e. tens of micrometres or more. The usefulness of the single-beam method was demonstrated by Luxmoore, Amin and Evans,[41] who used a mismatch technique to increase the sensitivity to fractions of a micrometre, whilst at the same time retaining the advantages of its greater robustness and simplicity over the two-beam method.

The single-beam method has proved the most useful in practical strain measurements,[41,46–48] and is recommended for such work. The two-beam method is more sensitive, and because it produces intensity changes in the individual speckles, it lends itself to electro-optical measurements. Butters[48,49] has developed a very impressive television system for scanning the specimen, storing the speckle intensity changes, and then reproducing the fringe patterns on a monitor. Future developments in this area could have considerable commercial implications (the system is currently marketed by Loughborough Consultants) but the hardware is expensive.

Use of Replicas

The speckle photographs, used to produce the interference fringes from which in-plane displacements are measured, can be replaced by replicas of the specimen surface, taken before and after loading. This procedure requires the specimen surface to have a fairly fine matt finish, so that successive identical replicas can be made. This limits the technique mainly to metal surfaces, but these form the major proportion of critically stressed structures. Replicas can be taken in difficult locations where it may be difficult to site a camera and laser, and they also avoid the location problems which lead to the geometric errors previously outlined.

The scattering characteristics of replicas should be almost identical to those of the specimen, so that, in principle, more information is available than from a speckle photograph. The writer and his colleagues have developed recently a transparent replicating system suitable for this purpose,[21,22] and Young's fringes can be observed by imaging one replica onto the other using a telecentric lens.[24] The replica only covers small areas

($\sim$ 50 mm square), but this is more than adequate for the study of stress and strain gradients around holes, cracks and across welds. Strain sensitivities as high as 10^{-6} are possible, and strain gradients can be plotted across distances as small as 1 mm.

Measurement of Surface Tilt

When the speckle pattern produced in front of the diffuser is observed on a screen it can be shown that if the diffuser rotates about an axis in its plane, i.e. tilts, the speckle pattern will also rotate, and produce a corresponding movement of the speckle pattern on the screen. The same effect can be obtained if the screen is placed behind a lens, provided the screen is not in the image plane, where the speckle pattern is virtually independent of diffuser tilt. Similarly, there is a position behind the lens where the speckle pattern is independent of in-plane movements of the diffuser (or specimen) and only sensitive to specimen tilt. For a specimen illuminated with a parallel beam of light, this 'tilt-only' position coincides with the back focal plane of the lens,[50] Fig. 6(a) but for the more usual case of diverging beam illumination, Fig. 18(b), the 'tilt-only' plane may be found by placing a mirror on the specimen surface and locating the image of the point source illumination behind the lens.[51]

A film placed in this position can record the speckle pattern, before and after tilt of the specimen, and after processing, the film can be viewed so as to produce Young's fringes or, if the tilt varies across the specimen surface, deformation fringes. The displacement d produced by the tilt can be measured from the Young's fringe spacing, and this displacement is related to the tilt γ by the equation

$$\gamma = \frac{(a + b - L)}{L(a + b)(1 + \cos\beta)} \times d$$

where a = distance of point source to specimen,
b = distance of specimen to lens,
L = focal length of lens, and
β = angle between illumination and camera axis (see Fig. 6(b)).

For parallel illumination, this equation reduces to

$$\gamma = d/L(1 + \cos\beta)$$

The most sensitive optical system requires $\beta \rightarrow 0$ and L to be as large as possible, although the latter is limited by practical considerations. Once these parameters are fixed, the sensitivity depends on the measurement of the displacement d. For typical optical arrangements, tilts of at least 10^{-6}

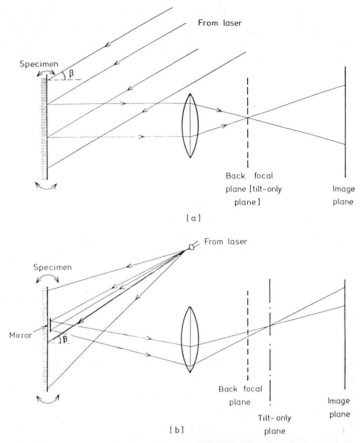

FIG. 18. Optical systems for separating speckle tilt effect from surface displacements (a) using collimated illumination (Tiziani) (b) using diverging illumination (Gregory).

radians can be measured. Unfortunately, the speckle in the tilt plane is sensitive to rotations of the specimen about an axis normal to the surface, and these must be separated from the tilt measurements.[51]

COMPARISON OF MOIRÉ AND SPECKLE SENSITIVITIES

At first sight, it might appear that the moiré and laser techniques are quite different, both in concept and practice. Though the latter is obviously so, both techniques depend on optical interferometry and diffraction, and conceptually they are very similar.[52]

The moiré gratings diffract light into discrete directions which are related to the pitch by the grating equation. For normal illumination of the grating, this can be written as

$$p \sin \theta_n = n\lambda \qquad (2)$$

where θ is the angle of diffraction of the nth order. The first order diffraction is the only one necessary to reproduce the main grating pitch, so putting $\sin \theta_n \simeq \theta$ for $n = 1$ we have

$$p = \lambda/\theta \qquad (3)$$

The relation between mean strain ε and fringe spacing f then becomes

$$\varepsilon = p/f = \lambda/f\theta \qquad (4)$$

If the zero order is stopped out, and only the $+1$ and -1 orders recombined, we effectively double the pitch of the grating[3] and the relation becomes

$$\varepsilon = \lambda/2f\theta \qquad (5)$$

For the double beam speckle method[42] the strain is given by

$$\varepsilon = \lambda/2f \sin \theta$$

$$\simeq \lambda/2f\theta$$

For small angles of illumination, this same equation also applies to Duffy's method.[44]

The sensitivity of the single-beam method is determined by the angle θ at which scattered light is observed from the speckle negative[21,41] and this can be written as object displacement,

$$d = mn\lambda/\sin \theta \simeq mn\lambda/\theta \qquad (6)$$

where m is the object/image ratio of the recording camera. For fringes n and $(n + 1)$, the difference in displacement is

$$d_{n+1} - d_n = m\lambda/\theta$$

but

$$\varepsilon = (d_{n+1} - d_n)/mf$$

and so

$$\varepsilon = \lambda/f\theta \qquad (7)$$

The similarity between all these equations is a consequence of the basic optical process of diffraction and interference used by all these techniques, and shows that greater sensitivity can always be achieved by increasing the effective scattering angle, θ.

REFERENCES

1. LUXMOORE, A. R., *Strain*, 1969, **5**, 1.
2. THEOCARIS, P. S., *Moiré Fringes in Strain Analysis*, Pergamon Press, Oxford, 1969.
3. DURELLI, A. J. and PARKS, V. J., *Moiré Analysis of Strain*, Prentice Hall, New Jersey, 1970.
4. LUXMOORE, A. R., *Exp. Mech.*, 1972, **12**, 216.
5. LONGHURST, R. S., *Geometrical and Physical Optics*, Longman, Green and Co., London, 1964.
6. GUILD, J., *The Interference Systems of Crossed Diffraction Gratings*, Oxford University Press, Oxford, 1956.
7. POST, D., *Exp. Mech.*, 1968, **8**, 63.
8. BOONE, P. and van BEECK, W., *Strain*, 1970, **6**, 14.
9. HOLISTER, G. S. and LUXMOORE, A. R., *Exp. Mech.*, 1968, **8**, 210.
10. COOK, R. W. E., *Optics Laser Tech.*, 1971, **3**, 71.
11. LUXMOORE, A. R., *J. Strain Anal.*, 1970, **5**, 162.
12. WADSWORTH, N. J., MARCHANT, M. J. N. and BILLING, B. F., *Optics Laser Tech.*, 1973, **5**, 119.
13. BOONE, P., *Proc. Int. Symp. Holography*, Besançon, 1970, Paper 5–1. International Commission of Optics.
14. LUXMOORE, A. R. and HERMANN, R., *Strain*, 1970, **6**, 115.
15. LUXMOORE, A. R. and HERMANN, R., *Exp. Mech.*, 1971, **11**, 375.
16. FESSLER, H. and THORPE, T. E., *Strain*, 1972, **8**, 14.
17. de CALUWÉ, M., BOONE, P. and van BEECK, W., *Strain*, 1971, **7**, 15.
18. LUXMOORE, A. R., HOUSE, C. and JONES, L. R., *J. Phys. E. (Sci. Instr.)*, 1971, **4**, 882.
19. LUXMOORE, A. R., *Proc. Conf. Recent Advances in Stress Analysis*, Roy. Aero. Soc., London, 1968.
20. CLARK, J. A., DURELLI, A. J. and PARKS, V. J., *J. Strain Anal.*, 1971, **6**, 134.
21. LUXMOORE, A. R., *Proc. Conf. Measurement in Civil Engineering*, Newcastle, 1977, Brit. Soc. Strain Measurement.
22. AMIN, F. A. A. and LUXMOORE, A. R., *Proc. Laser* 78 (Engineers Digest Technical Conference), London, 1978. Engineers Digest Limited.
23. LUXMOORE, A. R., *Optical Instruments and Techniques* (Home Dickson, J., ed.), Oriel Press, Newcastle upon Tyne, 1970.
24. LUXMOORE, A. R. and HOUSE, C., *J. Phys. E. (Sci. Instr.)*, 1972, **5**, 488.
25. BURCH, J. C. and FORNO, C., *Opt. Engineering*, 1975, **2**, 178.
26. FORNO, C., *Proc. Conf. Measurement in Civil Engineering*, Newcastle, 1977, Brit. Soc. Strain Measurement.

27. LEHMAN, M., *Holography: Technique and Practice*, Focal Press London, 1970.
28. HAINES, K. A. and HILDEBRAND, B. P., *Appl. Optics*, 1966, **5**, 595.
29. ENNOS, A. E., *J. Phys. E. (Sci. Instr.)*, 1968, **1**, 731.
30. BEES, T. J., B.Sc. thesis, 1972, Civil Engineering Department, University of Wales, Swansea.
31. GATES, J. W. C., *Optics Tech.*, 1969, **1**, 247.
32. TSUJIUCHI, J., TAKEYA, N. and MATSUDA, K., *Opt. Acta*, 1969, **16**, 709.
33. BOONE, P. M. and DE BACKER, L. C., *Optik*, 1973, **37**, 61.
34. LIGHT, M. F. and LUXMOORE, A. R., *Mag. Concr. Res.* 1972, **24**, 167.
35. ARCHBOLD, E., *Proc. Conf. Engineering Uses of Coherent Optics* (Robertson, E. R., ed.), Cambridge University Press, Cambridge, 1976.
36. BUTTERS, J. N., *Proc. Conf. Engineering Uses of Holography* (Robertson, E. R. and Harvey, J. M. eds.), Cambridge University Press, Cambridge, 1970.
37. LUXMOORE, A. R. and HOUSE, C., *Proc. Int. Symp. Holography*, Besançon, 1970, Paper 5–2 (International Commission of Optics).
38. BOONE, P. M., *Optics Laser Tech.*, 1970, **2**, 94.
39. LEENDERTZ, J. A., *Optical Instruments and Technology* (Home Dickson, J., ed.), Oriel Press, Newcastle upon Tyne, 1970.
40. ELLIASSON, B. and MOTTIER, F. M., *J. Opt. Soc. Amer.*, 1971, **61**, 559.
41. LUXMOORE, A. R., AMIN, F. A. A. and EVANS, W. T., *J. Strain Anal.*, 1974, **9**, 26.
42. ARCHBOLD, E., BURCH, J. M. and ENNOS, A. E., *Opt. Acta*, 1970, **17**, 883.
43. BUTTERS, J. N. and LEENDERTZ, J. A., *J. Phys. E. (Sci. Instr.)*, 1971, **4**, 277.
44. DUFFY, D. E., *Appl. Optics*, 1972, **11**, 1778.
45. ARCHBOLD, E. and ENNOS, A. E., *Opt. Acta*, 1972, **19**, 253.
46. EVANS, W. T. and LUXMOORE, A. R., *Eng. Fract. Mech.*, 1974, **6**, 735.
47. AMIN, F. A. A. and LUXMOORE, A. R., *J. Inst. Metals*, 1973, **101**, 208.
48. BUTTERS, J. N. and LEENDERTZ, J. A., *J. Measurement and Control*, 1971, **4**, 349.
49. BUTTERS, J. N., *Proc. Conf. Engineering Uses of Coherent Optics* (Robertson, E. R., ed.), Cambridge University Press, Cambridge, 1976.
50. TIZIANI, H. J., *Optics Comm.*, 1972, **5**, 271.
51. GREGORY, D. A., *Proc. Conf. Engineering Uses of Coherent Optics* (Robertson, E. R., ed.), Cambridge University Press, Cambridge, 1976.
52. LUXMOORE, A. R. To be published.

INDEX